D0554420

THE DISCOVERY OF NEPTUNE

THE DISCOVERY OF
NEPTUNE

• • • • • • • • • • • •

Morton Grosser

HARVARD UNIVERSITY PRESS
Cambridge, Massachusetts · 1962

LIBRARY
SEP 26 1962
UNIVERSITY OF THE PACIFIC

© Copyright 1962 by the President and Fellows of Harvard College

All rights reserved

•

Distributed in Great Britain by Oxford University Press, London

Library of Congress Catalog Card Number: 62–17218

Printed in the United States of America

112431

ACKNOWLEDGMENT

I would like to express my thanks for the use of documents in the library of St. John's College of Cambridge University and the Library of the Paris Observatory. Mr. F. E. Brasch, curator of the Newton Collection of the Stanford University Library, was a gracious source of encouragement to me throughout this work. Mr. Jack McDowell and Professors David Harris, Frederick O. Koenig, and Charles Donald O'Malley read the manuscript and made many helpful suggestions. My most indispensable helper has been my wife Janet, to whom this book is dedicated.

M. G.

CONTENTS

THE DISCOVERY OF NEPTUNE

INTRODUCTION

• • • • • • • • • • • •

A Brief Historical Survey of Planetary Astronomy to 1781

DISTINGUISHED by their motion against the background of fixed stars, the planets were familiar objects to Babylonian and Egyptian astrologers. Long before the sixth century B.C. the Egyptians gave the wandering stars euphonious names (Doshiri = Mars) and were familiar with individual peculiarities of their movements, including retrograde motion. The Greek philosophers, beginning with Eudoxus of Knidus (408?–355? B.C.), and including Hipparchus (*fl.* 150 B.C.) and Ptolemy of Alexandria (*fl.* A.D. 150), attempted to systematize the planetary movements mathematically; the Ptolemaic solution became a standard point of departure for attacking this difficult problem.

Ptolemy's system ranged the heavenly bodies around a central Earth in this order of radial distance: Moon, Mercury, Venus, Sun, Mars, Jupiter, Saturn. It was, in a strict mathematical sense, a planetary construct, an attempt to describe complex movements using only components of circular motion. Eventually it was replaced by the system of Nicolaus Copernicus (1473–1543), who interchanged Ptolemy's positions for the Earth and the Sun and attempted some simplification. So similar are the geometrical idioms of the Ptolemaic and Copernican solutions that the latter may be regarded as the most successful (but not the last) of the Ptolemaic revisions which preoccupied Western astronomers for fifteen centuries.

During most of this time the planets could be identified only by their changing positions in the heavens. These positions were determined with increasing precision as the techniques and instruments used for line-of-sight measurements were improved. Tycho Brahe (1546–1601) brought naked-eye observation to unprecedented heights of accuracy; the coordinates of the standard stars in his catalogue were defined with probable errors of less than a minute of arc. Tycho's observational records of the planet Mars were bequeathed to his pupil, Johann Kepler (1571–1630), in 1601. Working with these data, Kepler formulated, after five years of trial-and-error calculation, his first two laws of planetary motion. The laws, which were published in the *Astronomia nova* in 1609, stated that: (1) the planet Mars describes an elliptical orbit, with the Sun at one focus of the ellipse; (2) the radius vector from the Sun to Mars sweeps out equal areas in equal times. These statements provided the first fundamentally accurate description of planetary orbits. Their departure from the historically dominant credo of circular motion marks a major turning point in descriptive astronomy.

A few months after the publication of *Astronomia nova* there occurred the greatest single advance in the history of observational technique: Galileo Galilei (1564–1642) turned his own version of the Dutch optician Lippershey's spyglass to the night sky over Padua and saw "great, unusual and remarkable spectacles" such as no man had seen before. He described these sights in a now-famous little book, the *Siderius nuncius,* published at Venice in March 1610. Among its many revelations was a new discriminant for the planets:

> Deserving of notice also is the difference between the appearances of the planets and of the fixed stars. The planets show their globes perfectly round and definitely bounded, looking like little moons, spherical and flooded all over with light; the fixed stars are never seen to be bounded by a circular periphery, but have rather the aspect of blazes whose rays vibrate about them and scintillate a great deal.[1]

This fact, that planets appear through a telescope as discrete optical disks, while stars are seen, independent of magnification, as points of light, was a vital contribution to planetary identification. Together with many of Galileo's telescopic discoveries, it was confirmed by observations made by other astronomers during the next ten years.

Throughout this decade, Johann Kepler continued his laborious calculations from Tycho's data. Working in a unique discipline compounded of orbital mechanics, metaphysical geometry, and cosmological music, Kepler formulated, in May 1618, his celebrated third law of planetary motion. This statement, that the squares of the periods of revolution of any two planets are proportional to the cubes of their mean distances from the Sun, was a powerful and widely applicable relationship; it fitted not only the orbits of all the known planets, but those of the four newly discovered satellites of Jupiter with respect to their parent planet as well.

Kepler's discoveries defined, within the limits of contemporary observational accuracy, where the planets were, and, given their periods of revolution, where they could be. His three laws, which established the first correct principles of planetary mechanics, were potential replacements for a vast corpus of inaccurate theory. It took years for them to become even moderately well known. The inertia of entrenched viewpoints was partly responsible for this, but Kepler's form of presentation was even more to blame.

Readers shied away from his thick, difficult volumes, which were full of obscure mystical speculation. Before the publication of Isaac Newton's *Principia* in 1687, few men cared to winnow the minute, if rich, scientific yield from Kepler's immense storehouse of metaphysical chaff.

One exception was a young English astronomer, Jeremiah Horrocks (1619–1641). Horrocks studied Kepler's planetary theory and postulated an elliptical lunar orbit, thus anticipating the work of some outstanding physicists and astronomers who flourished in the late seventeenth century. During Horrocks's lifetime and for some time after, scientific investigation of the planets proceeded along two fairly discrete fronts: theoretical mechanics and telescopic discovery. The latter field, which was still in its infancy, produced sensation after sensation.

Galileo's harvest was perhaps the richest. In addition to the optical differentiation of planets from stars, he described the phases of Venus, the physical features of the Moon, four satellites of Jupiter (the "Medicean stars"), and, although he was unable to resolve it completely, the unusual appearance of Saturn. Galileo's investigation of sunspots is too well known to detail here; although the Jesuit astronomer Christopher Scheiner (1575–1650) announced the discovery first, Galileo's analysis of the phenomenon far surpassed any previous efforts. Father Scheiner was one of many diligent searchers. His adoption of the "Keplerian" common-focus astronomical telescope in 1618 presaged the rapid spread of telescopic observation throughout Europe. Though some of the instruments then used had in-focus cross hairs, even the best early Keplerian telescopes left much to be desired in optical performance. They were to receive major technical improvements from Christian Huygens.

Huygens (1629–1695) was born at The Hague, and

studied at Leiden and Breda. His early observations convinced him that the optical elements of contemporary telescopes were inadequate, and, working with his older brother, he experimented with methods of producing more accurate lenses. In 1655, using a glass which he had figured and polished himself, Huygens discovered Titan, the largest satellite of Saturn, and also solved the long-standing mystery of that planet's true configuration. He announced the solution, encoded as a Latin anagram, in *De Saturni luna observatio nova,* and after further study published in *Systema Saturnium* (1659) a more complete explanation of the various appearances of Saturn, correctly attributing them to the phases of an encircling ring inclined 20° to the ecliptic.

A contemporary of Huygens, Gian Domenico Cassini (1625–1712), also made careful observations of Saturn. His patience was rewarded when he discovered what is now known as the "Cassini division," a major dark separation within Saturn's rings. After studying the gap, Cassini deduced that (as presently believed) the rings were composed of swarms of satellites too small to be resolved individually, circling the planet with different velocities. Cassini also discovered Saturn's second-largest moon, Iapetus, and found that, like the Earth's moon, it always presented the same face to its central planet. The following year, in 1672, he found a third moon, Rhea, and twelve years later he discovered two more, Tethys and Dione.

Cassini did not confine his observations to the ringed planet. Most of his work on Saturn was done after his arrival at the Paris Observatory in 1669, but earlier in his career, while teaching astronomy at the University of Bologna, he had spent much time observing Venus, Mars, and Jupiter. In the process he discovered and attempted to measure the periodic rotation of the three planets about their axes. Cassini noted many surface features on both Mars and Jupiter. He described the visible Martian landscape, marking its seasonal changes, and compared its white polar caps with the terrestrial icecaps. In 1665, while attempting to map the markings on Jupiter's surface, he discovered the Great Red Spot, an area about 48,000 kilometers long and 11,200 kilometers wide that drifts across the planet's south temperate zone.

One of the problems that confronted Galileo after he had discovered the Medicean satellites of Jupiter was computing their orbits. He was never successful in this, and the problem was finally solved by Cassini, who published in his *Ephemerides Bononienses Mediceorum siderum* (1668) remarkably accurate tables for the motions of Io, Europa, Ganymede, and Callisto. The motion of the Earth was also of interest to Cassini. He worked on the eccentricity of the terrestrial orbit and the obliquity of the ecliptic, but most significant was his attempt to determine the Earth's exact distance from the Sun, which was inaccurately known and drastically underestimated before Cassini's time. In collaboration with Jean Richer (d. 1696), who made duplicate observations at Cayenne, French Guiana, Cassini measured the apparent distance of Mars from several fixed stars during the opposition of 1672. From the two sets of observations, made 7200 kilometers apart, it was possible to triangulate the position of Mars, and hence that of the Sun, with theretofore unequaled accuracy. Cassini estimated the solar parallax to be $9''.5$, corresponding to a distance from the Earth of 139.2×10^6 kilometers; this figure, published in 1731, is a few percent less than the contemporary accepted value (149.53×10^6 kilometers).

During this period of steady progress in descriptive astronomy the work of systematization seemed to make little headway. While it is clear that advances in planetary mechanics were more difficult and less sensational than telescopic discoveries, this does not entirely explain the lag. One of the reasons was undoubtedly the opposition of theologians. Neither Protestant nor Catholic clerics were particularly cordial to the Copernican scheme before any offical position was taken; after the papal edict of 1616, resistance to the heliocentric system was widespread even among the educated class of predominantly Catholic populations. Both Gian Cassini and René Descartes admitted that their public astronomical opinions were affected by the attitude of the Roman Church. The Church, however, did not damn all practical astronomy; it objected only to a Sun-centered planetary system, and, more particularly, to a moving Earth. It is difficult not to believe that such specific ecclesiastical prohibition was partly responsible for the disparity between the advancement of telescopic observation and that of planetary mechanics.

Certainly the method of advancement in the two areas was markedly different. Telescopic exploration was a modified, but not highly disciplined, trial-and-error process which differed little from the methods that had been used by European protoscientists since the Middle Ages. Each new discovery was the product of individual inquiry, was relatively independent of discoveries preceding it, and exerted only moderate pressure on those following it. Because of the wealth of previously unknown objects, visible only through a telescope, discoveries were frequent, and any anticipatory intellectual tension was relieved at short intervals. In contrast, the systematization of planetary motion progressed by large steps, often widely separated in time. Each advance was a revision of existing theories and required not only perseverance but also the rare ability to combine many elements into a unified construct. In the decades between major advances a high degree of tension built up, and, though this tended to focus attention on the critical problems, the problems themselves were so difficult that they could be solved only by men of extraordinary brilliance.

Kepler was well aware of the most important discrepancy in planetary theory. His three laws answered some of the questions about the planets, specifically, where they were and, in part, how they moved. He recognized that the crucial question was why they moved as they did, and he made an attempt to answer it. His solution combined a theory of planetary magnetism similar to that formulated by William Gilbert (1540–1603), with a mysterious primal force called the *anima motrix* which was assumed to reside in the Sun and to be selectively extensive in space. It was a bizarre and wholly qualitative system, and, even allowing the heroic rationalizations that Kepler made in its favor, it did not work. This was true of all the general theories of the solar system devised between Kepler's death and Isaac Newton's successful solution of the problem of planetary motion.

For any system of planetary mechanics to work, it had to save the phenomena, and thus be compatible with Kepler's laws. Some theorists who were unable to satisfy these conditions decided arbitrarily to alter or reject Kepler's laws. Chief among such men were Ismael Boulliaud (1605–1694), who devised a strange and conspicuously inaccurate arrangement of oblique cones, and Seth Ward (1617–1689), a Savilian professor at Oxford, whose planets moved with uniform angular motion around the unoccupied focus of their elliptical orbits. Equally unsatisfactory were the efforts of René Descartes (1596–1650) and Giovanni Alfonso Borelli (1608–1679) to construct general systems. Although each scheme embodied some ingenious ideas, both suffered from vague dynamic arrangements. Neither man offered any quantitative information about his system's motivating force, and neither one explained the observed phenomena with any precision. Nevertheless, despite its shortcomings, Descartes's system gained many adherents on the Continent because of its generality and the renown of its author.

While the Cartesians were speculating on cosmic vortices, a new theory of planetary motion was being evolved elsewhere. Its background was more mundane than the imaginative schemes of earlier cosmology; it originated in studies of the dynamics of earthly particles.

It has been said that the crucial turning point in planetary mechanics was the acceptance of a straight rather than a circular path as the "natural" one for the motion of a free body. If this is so, astronomy owes a heavy debt to Galileo for his experiments with falling bodies. Huygens applied the idea to the planets: assuming that the undisturbed orbit of both celestial and earthly bodies was a straight line, he calculated the radial force necessary to keep a planet in a circular path to be mv^2/r, where m is the mass, v the velocity, and r the orbital radius. He then postulated that for circular planetary orbits, saving the rest of Kepler's laws, the central force must vary as $1/r^2$. This conclusion was also reached by three British scientists. Christopher Wren (1632–1723), Robert Hooke (1635–1703), and Edmund Halley (1656–1742), all members of the newly founded Royal Society, were intensely interested in planetary motion. They knew of Kepler's work, and their studies led them to ask a question which they could not answer: Would a planet subject to a force varying inversely as the square of the distance describe an elliptical orbit compatible with Kepler's laws?

The mathematical techniques needed to solve this problem did not exist when Isaac Newton (1642–1727) attacked it in 1665. Working alone, he devised the necessary analytical tools and applied them with outstanding success to the mechanics of the solar system. Newton's brilliant mathematical talent was not his sole aid in solving a problem that had baffled many other minds. Perhaps equally important was his deep-seated conviction that celestial phenomena were subject to ordinary (that is, terrestrial) mechanical laws. Only a man who believed in a continuous realm of physical action extending from the Earth to the farthest reaches of space could have made the famous analogy between an apple and the Moon. Newton concluded from his early studies that three type-problems had to be solved in order to prove the existence of a universal force pervading the solar system. These were the determination of the motions of (1) the Sun and a planet (two points, one traveling in an elliptical orbit around the other); (2) the Earth and the Moon (two finite spheres); (3) the Earth and a small body near its surface. The last two cases involved the difficult analysis of gravitational attractions exerted by various parts of a body not of negligible size. At different times Newton was successful in solving all of these problems. His solutions (in modified form) were published together for the first time in *Philosophiae naturalis principia mathematica* (1687), possibly the most momentous scientific book ever printed, and certainly the greatest single contribution to planetary mechanics.

In the third section of the *Principia*, which was titled "De mundi systemate," Newton dealt with the motion of the planets. Beginning with the moons of Jupiter, he showed that bodies moving under the influence of a central attractive force conformed to Kepler's laws. By induction he extended these conclusions to the entire solar system, confirming both Kepler and Copernicus, and established the law of universal gravitation: the force between any two bodies is directly proportional to the product of their masses and inversely proportional to the square of their separation. Newton then made what has been called his master stroke. He showed that the single universal force defined by the foregoing relation determined the paths of the planets around the Sun, the orbits of satellites around the planets, the behavior of terrestrial falling bodies, and the movements of the tides.

This magnificent synthesis, unprecedented in breadth and power, was extended even further. In the same section of the *Principia* Newton laid the foundations for the study of comets, treating them as planets with highly eccentric orbits, and followed this with equally novel and brilliant conclusions about the figure of the Earth.

Newton emphasized one concept among the revelations of the *Principia* that was to play a vital part in the technique of planetary discovery. This was the idea that every particle, however small, acted on every other particle with an effect proportional to the product of their masses. Theoretically, this doctrine of mutual interactions transformed Kepler's simple planetary motions into the complex resultants of uncounted individual forces. In practice, however, limited observational accuracy precluded detection of all but the largest effects. It is important to realize that, though the principle was unchanged, the degree of its application depended on the technology of astronomical instruments. For any reduction in the tolerances of angular measurement there would follow a corresponding reduction in the magnitude of the smallest detectable perturbation effect. There was such an improvement in measurement; the practical result of the above-mentioned proportionality was that, in time, gravitational effects of smaller bodies or bodies farther away, or both, on the members of the solar system could be analyzed. Thus Newton's work closely united the various pursuits within astronomy. For the first time, theory could receive immediate and direct benefit from improved observational accuracy.

By a fortunate coincidence, at the end of the seventeenth century there was an urgent commercial demand for more accurate observations. The motivation was the need for a reliable way of determining longitude. During the preceding century several European governments had offered rich prizes for a successful method, but none had been found. By the reign of Charles II of England, the rapidly increasing frequency of east–west trade voyages and the expansion of the British navy had made the determination of longitude a pressing royal concern. One result of the crown's anxiety was the founding of the Greenwich observatory, and the concurrent appointment of John Flamsteed (1646–1719) as the first Astronomer Royal (1675).

Flamsteed was an excellent choice. He was both industrious and convinced of the need for more accurate star and planet observations. At great cost to himself of funds and labor he equipped the observatory and spent years making the observations which were eventually compiled into the *Historia coelestis Britannica* (published in its final form in 1725). This catalogue, which gave the positions of 2935 stars, was more accurate than any previous work by an entire order of magnitude. (Although some of Tycho Brahe's observations were, as mentioned earlier, extremely fine, his average error was 4′. Flamsteed's positions were all accurate to 10″ of arc.)

Flamsteed was succeeded at Greenwich by Edmund Halley, who began building the great astronomical superstructure that eventually stood on Newton's foundation. Applying the methods of the *Principia,* Halley determined the orbits of twenty-four comets and discovered the periodicity in the appearances of the one that bears his name. He also recognized the possibility of utilizing the next transit of Venus, which he predicted would occur in 1761, for measuring the solar parallax. Although he did not live to observe it, he left specific instructions for the eighteenth-century astronomers who survived him.

The century following the publication of the *Principia* produced important advances in celestial mechanics. Newton's ideas were extended and refined by men who were (as they now had to be) primarily brilliant mathematicians. Many of them studied the gravitational interactions of celestial bodies (perturbation theory), and a remarkable number of them were French. One of their preoccupations was the attempt to determine the motions of three mutually interacting bodies. All efforts to emulate Newton, who had completely solved the problem for two bodies, by finding an exact general solution failed. We now know that the general form of the problem is insoluble in terms of known analytic functions. Approximate solutions were found by three outstanding mathematicians: Leonhard Euler (1707–1783), Alexis Claude Clairaut (1713–1765), and Jean Le Rond d'Alembert (1717–1783).

Euler was particularly interested in the Moon's motion, and it was in connection with this that he attempted to solve the three-body problem. His improvements in lunar theory were published at Saint Petersburg in the *Theoria motus lunae* (1753). This theoretical work enabled Tobias Mayer (1723–1762), a German astronomer, to construct the first lunar tables accurate enough for the determination of longitude at sea (1755). Euler also made significant contributions to gravitational theory and recognized its disconcerting implications. Allowing perturbations, the orbit of a planet could no longer be regarded as a simple Sun-focused ellipse; it also reflected the dynamic histories of all other bodies affecting it at any instant. The elements of a planet's path were therefore different at each epoch, depending on the position of other planets. Euler attempted to solve this problem by considering two bodies at a time and assuming that one of them was moving, instantaneously, in an elliptical orbit whose elements were being continuously altered by the action of the second perturbing body. He was able, given the position of the perturbing body, to determine the elements of the successive ellipses and predict the position of the first planet.

Clairaut worked primarily on the figure of the Earth, but he made some interesting contributions to celestial mechanics as well. Knowing of Halley's prediction that the comet of 1682 would return in 1758, Clairaut calculated the perturbations of the comet caused by the major planets near its path (Jupiter and Saturn) and attempted to predict the exact date of its perihelion. His result, April 13, 1759, was given with a tolerance of ± 1 month. The comet was seen again on Christmas day, 1758, confirming Halley, and reached perihelion on March 13, 1759, missing Clairaut's earliest prediction by 1 day. Clairaut also published work on the position of the Moon. He based his calculations on the observations made by Nicolas-Louis de La Caille (1713–1762), and was remarkably successful in describing the real lunar motion. Clairaut and Euler differed with d'Alembert on the nature of the moon's motion, but the constant interchange between the three men sharpened all their ideas, and, perhaps as a result, they each published significant works on lunar theory.

One of the puzzling components of the Moon's motion was libration, a small apparent oscillation in latitude and longitude. The Paris Academy offered a prize for research on the problem, and in 1764 it was won by a young Franco-Italian mathematician, Joseph Louis Lagrange (1736–1813). Besides his prize essay on the Moon Lagrange published the first analysis of the dynamics of Jupiter's satellites, and discovered the invariability of the mean solar-planetary distances. It has been suggested, however, that his most important service to astronomy was in supplying analytic methods of great power and generality to his brilliant contemporary, Pierre Simon, Marquis de Laplace (1749–1827).

Although Laplace was born before mid-century, his major work, the *Mécanique céleste,* did not begin to appear until 1799. From 1773 to 1784 he worked on the stability of the solar system. Using techniques developed by Euler and Lagrange, he showed that the mutual attractions of Jupiter and Saturn cause periodic and not cumulative inequalities in their mean motions. Extending this investigation, Laplace found that mutual secular perturbations are periodic and apparently stable for the mean motions and distances of all the planets, and working with Lagrange, he later proved that both the inclinations and eccentricities of the planetary orbits were free to vary only within narrow limits. As the last quarter of the eighteenth century began, Europeans could be confident that the celestial world, at least, would remain stable in the foreseeable future.

At this time Englishmen had more than one reason to be skeptical of terrestrial stability. In addition to their political woes, they had been informed some years earlier by the third Astronomer Royal of a new addition to the Earth's repertoire of motions. James Bradley (1692–1762) followed in the tradition of his predecessors at Greenwich. After making extremely careful observations, he announced in 1748 that he had discovered a small periodic oscillation of the Earth's axis, a second-order inequality in the motion of precession. Bradley attributed it to the attraction of the Moon on the Earth's equatorial bulge.

He described this nutation as a retrograde elliptical movement of the pole about its mean position; the dimensions of the ellipse were 18″ by 16″, and the period of motion was 18 years. In a word, the Earth wobbled.

Nutation was not Bradley's first important discovery. Earlier, by repeated near-vertical observations of the star γ Draconis, he had found small periodic variations in position which he eventually determined were due to the aberration of light. It was the further investigation of this phenomenon that ultimately led to the discovery of the Earth's nutation.

Bradley's successor Nevil Maskelyne (1732–1811), who was appointed in 1765, continued the high quality of astronomical inquiry that the scientific world had come to expect of Greenwich. One of Maskelyne's accomplishments was correlating and reducing the many widely scattered observations of the 1769 transit of Venus. Not long after 1769, however, the work of the Royal Observatory was overshadowed by astronomical developments elsewhere in England. Professional observers of all nationalities turned their attention to the remarkable discoveries of a naturalized British amateur, an itinerant musician from Hanover who had taught himself mathematical astronomy at thirty-five years of age.

CHAPTER 1

• • • • • • • • • • • • • •

URANUS, 1781–1809

WILLIAM Herschel was born in 1738, the second son of an oboist in the Hanoverian foot-guards band. Although his vocational study was music, he was instructed by his father in a wide variety of other subjects, including rudimentary astronomy. Herschel joined Hanover's army at fourteen years of age, but after a short unhappy career as a military bandsman he decided to leave for England, and arrived there in November 1757. He quickly found employment as a skilled musician, but an intense intellectual curiosity dictated his leisure pursuits. He studied first languages (English, Italian, Latin, and Greek, in that order) and then, by way of harmonic theory, mathematics. In the meantime his musical reputation and income increased steadily.

Herschel's mathematical studies also progressed rapidly, and in 1773 they brought him to a turning point in his life. By that time it was his pleasure to

> unbend the mind (if it may be so called) with a few propositions in *Maclaurin's Fluxions* or other books of that sort.
>
> Among other mathematical subjects, Optics and astronomy came in turn, and when I had read of the many charming discoveries that had been made by means of the telescope, I was so delighted with the subject that I wished to see the heavens and planets with my own eyes thro' one of those instruments.[1]

This first attraction was the beginning of a lifelong pursuit. Working in the leisure left from a busy musical career, with tremendous industry and enthusiasm, Herschel began to build his own telescopes. After experimenting with refractors he turned to reflecting instruments, and spent five years developing superior methods of grinding and polishing parabolic mirrors. The improved telescopes he constructed, which were, at that time, better than any at the Royal Observatory, were the means to an end.

Herschel's ambition was to measure the distribution of stars in space and thus to extend the knowledge of orderly cosmic structure from the solar system to the entire stellar universe. Aware that no existing celestial atlas was even remotely adequate for this project, he decided to remedy the deficiency singlehanded. The stupendous task he set himself as a life work was a complete and systematic survey of the heavens, a plan that required telescopic scrutiny of every visible object, a description of its appearance and position, and subsequent checks to verify its coordinates. In 1781 Herschel set to work with his favorite reflector, which had an aperture of 6.2 inches and a focal length of 7 feet, each night sweeping a measured section of the sky.

One of Herschel's subsidiary objectives was to determine the distances of some fixed stars. He was attracted to Galileo's suggestion for detecting stellar parallax by measuring periodic changes in the separation of close pairs of stars differing in brightness, and he began a methodical search for suitable star pairs.[2] In the course of this search, on the night of March 13, 1781, he noticed a curious object in the neighborhood of H Geminorum. Increasing the power of his eyepiece he found that the object had a magnifiable disk; continued observation convinced him that it was in motion relative to nearby stars. Although he did not realize it at the time, Herschel had discovered the seventh planet.

The new addition to the solar system was troublesome

from the moment of its discovery. In his first manuscript notation concerning it Herschel wrote that he had seen "a curious either nebulous star or perhaps a comet." [3] A few nights after he had first observed it he "found that it is a comet, for it has changed its place." The possibility that his discovery was a planet, though perfectly compatible with the visual evidence, did not occur to Herschel initially. His lapse was understandable. For more than 2000 years, as far as anyone knew, there had been only six planets; their number had remained constant despite frequent disagreements about their size, distance, and arrangement. Of the three men to whom Herschel first communicated the news of his discovery, only Nevil Maskelyne, the Astronomer Royal, preserved a sufficiently open mind to recognize that the object might be a planet. He mentioned this possibility in a letter written on April 4, 1781 to William Watson, Jr., who had acted as an intermediary between Herschel and the Astronomer Royal, and on April 23 Maskelyne wrote to Herschel himself:

> I am to acknowledge my obligation to you for the communication of your discovery of the present comet, or planet, I don't know which to call it. It is as likely to be a regular planet moving in an orbit nearly circular round the sun as a Comet moving in a very eccentric ellipsis.[4]

Nevertheless, Herschel's paper on his discovery, read before the Royal Society on April 26, was entitled simply "Account of a Comet," and it was some time before the astronomers of Europe were able to convince themselves otherwise.[5] They were not helped by Herschel's incorrect statement (read but not printed) that the supposed comet's daily parallax was between 10″ and 20″ and its position was therefore within the Earth's orbit.

Many astronomers on the continent, notified of the discovery by Maskelyne, attempted to determine the motion of the new body. Charles Messier (1730–1817) began his observations as early as April 16, 1781; he was followed by Joseph Jérôme Le Français de Lalande (1732–1807), Pierre François André Méchain (1744–1804), and Johann Elert Bode (1747–1826), to name only a few. At Paris, several attempts were made to compute a suitable parabolic (cometary) orbit as soon as a few observations had been accumulated. These efforts were unsuccessful; each set of elements coincided with the observed positions for a few days and then rapidly diverged. The wasted work and frustration can be charged to a single cause: an unquestioning acceptance of Herschel's discovery as a comet, despite the fact that the visual evidence contradicted such a conclusion.

In the note to Watson mentioned above, Maskelyne wrote that he had observed the object described by Herschel and that he was convinced that it was "a comet or new planet, but very different from any comet I ever read any description of or saw." In a note to Herschel dated April 29, 1781, Messier wrote that Maskelyne having informed him of

> a new comet, of very singular character, having no atmosphere, no coma nor tail, resembling a little planet with a diameter of 4 to 5 seconds, a whitish light like that of Jupiter, and the appearance when seen with glasses of a star of the 6th magnitude, I sought it . . . and I found that it had the characters which Mr. Maskelyne noticed.
>
> I am constantly astonished at this comet, which has none of the distinctive characters of comets, as it does not resemble any one of those I have observed, whose number is eighteen.[6]

Clearly, for those whose opinions were guided by their eyes alone, the classification of the new comet was open to question. Others, accepting the untrustworthy astronomical evidence of their ears, spent tedious hours calculating orbits inherently doomed to fail.

The erroneous assumption that the object was a comet carried with it a number of implications, the most important of which was that its perihelion would probably lie within the Earth's orbit and certainly not fall more than four astronomical units beyond it. This probability was very high. Of the many comets observed before 1781, only that of 1729 had shown a perihelion of more than four times the radius of the terrestrial orbit. The incompatibility of such a constraint with the movements of Herschel's object (his remarks on parallax notwithstanding) was not realized until Laplace, Méchain, and others had repeatedly failed to reconcile the cometary hypothesis with the observations. On May 8, 1781, Jean Baptiste Gaspard Bouchart de Saron (1730–1794) announced that the "comet" was really much more remote from the Sun than had been supposed, having a minimum perihelion twelve times the terrestrial-orbit radius. Not long after this the true nature of Herschel's discovery became apparent.

Anders Johann Lexell (1740–1784), astronomer at Saint Petersburg, was in London during the spring and summer of 1781. Summing up the characteristics of the new object, including its steady light, its lack of nebulosity, its slow motion in latitude, and its movement along the zodiac, he concluded that it was a planet, and on that basis he computed an orbit. Attempting to satisfy two observations made respectively by Herschel on March 17 and Maskelyne on May 11, 1781, Lexell calculated the first circular elements of the new planet's orbit. He determined its mean radius as 18.93 astronomical units, but later found that many different solutions satisfied the short observed path thus far traversed by the slow-moving body. Other sets of circular elements were published soon after in Russia, France, and Germany, but by then it was obvious that many more observations made over a long period would be required before the orbit could be represented with any accuracy.

The idea that a new planet had been discovered was widely accepted after Lexell announced his results. Both German and French journals printed acknowledgements of the fact; in the *Mémoires* of the French Academy of Sciences (1779) Lalande wrote, "It seems to me that we must refer to [Herschel's discovery] as the new planet." The magnitude of Herschel's unique accomplishment was suddenly recognized outside astronomical circles. In July 1782 he was appointed court astronomer to George III of England, and soon after he was called upon to propose a name for the new member of the solar system.

The naming of the seventh planet was a confused process. Sir Joseph Banks (1743–1820), then president of the Royal Society, first broached the subject in a letter written to Herschel on November 15, 1781. It was the end of the following year before Herschel finally suggested, out of gratitude to his royal patron, the awkward appellation "Georgium Sidus." Several critics pointed out that this name was misleading as well as cumbersome; "Sidus" referred to a star, which Herschel's discovery was not. The name was soon altered to "the Georgian Planet," or simply "the Georgian," which became standard British usage. Meanwhile, on the continent, the desk of Johann Bode, editor of the *Berliner Astronomisches Jahrbuch,* had become a central clearing house for other suggested names. From Upsala, Professor Erik Prosperin (1739–1803) submitted "Neptune" (!). In Russia, Lexell concurred with Prosperin and suggested to the Saint Petersburg Academy the names "Neptune de George III" and "Neptune de Grande-Bretagne." Many other mythological names were put forth, including "Minerva," "Cybele," and "Austräa," but Bode himself made the most appropriate suggestion. He pointed out that if the new planet was named "Uranus" the solar system would represent a coherent mythological family: Uranus, the god of the sky and husband of Earth, was the father of Saturn and the grandfather of Jupiter, who, in turn, fathered Mars, Venus, Mercury, and Apollo (or the Sun). Other European astronomers recognized the aptness of Bode's proposal, and Abbé Maximilian Hell (1720–1792), the Jesuit director of the Vienna Observatory, adopted the name "Uranus" in the *Viennese ephemerides* and published a Latin poem urging its acceptance. In France, however, the writings of Joseph Lalande had already swayed public opinion in favor of adopting his suggested name, "Herschel." Although in time "Uranus" gained some ground at the expense of Lalande's "Herschel," neither the French nor the English capitulated officially, with the result that three different names and two symbols for the same planet were used simultaneously for sixty years after its discovery.

A wide choice of angular diameters was also available for the planet. Herschel's first twelve measurements, all made in the spring of 1781, showed a variation of over 100 percent; the values ranged from 2″.53 to 5″.20. The average, 4″.04, was larger than Maskelyne's 3″, almost identical with Lexell's 4″, and considerably less than the 7″ estimated at Milan. Herschel later attempted to reach a more precise result with a new micrometer of his own design, making most of the observations during October 1782. This second series of ten measurements varied from 3″.51 to 5″.11 and averaged 4″.17. Using an angular diameter of 4″, and Lalande's value of 18.93 astronomical units for the mean radial distance of Uranus, Herschel computed the ratio of the new planet's diameter to that of the Earth as 4.454. (The ratio used at present is approximately 3.69.)

In the next few years many astronomers tried to replace the early circular orbits with more accurate descriptions of the motion of Uranus. Laplace and Méchain collaborated on the calculation of the first elliptical elements, which were published in 1783, and tables of the new planet appeared in the French and German astronomical ephemerides for 1787 (published in 1784). All of this work was necessarily based on positions observed during the preceding two years.

The short period of observation was a severe handicap to astronomers trying to compute the orbit of a planet moving as slowly as Uranus; Johann Bode realized that this limitation might be overcome. Hoping that earlier astronomers might have recorded Uranus as a star, Bode began to search through old astronomical catalogues and was rewarded almost immediately. In August 1781 he discovered an anomalous observation in the star catalogue compiled by Tobias Mayer (1723–1762) of Göttingen. On September 25, 1756, Mayer had listed a star (his number 964) in Aquarius, having a right ascension of 348° 0′ 20″ and a declination of 6° 2′ 3″. Bode found that the star was not visible at this position in 1781, and that Mayer had never again mentioned it on the subsequent occasions when he had observed all stars of comparable magnitude in the same vicinity. Using the orbits published by Laplace and Méchain, Bode calculated that in September 1756 Uranus would have been in nearly the same place as Mayer's number 964; thus it became highly probable that the object which Mayer had regarded as a fixed star was actually Uranus.

Continuing his search with the help of Placidus Fixlmillner of Kremsmünster (1721–1791), Bode found, in 1784, that Flamsteed had also observed Uranus, alias 34 Tauri, on December 23, 1690. That observation was discovered and verified independently by the French astronomers Lemonnier and Montaigne. It was recognized that, though Uranus had a very long period (84 years 7 days), the available observations now spanned more than one complete revolution and it should therefore be possible to compute an accurate orbit. Using the positions of 1690 and 1756 with the oppositions of 1781 and 1783, Fixlmillner calculated the following elliptic elements, which satisfied all the observations known to that time: [7]

Epoch, January 1, 1784

Mean anomaly	297°	9′	25″
Longitude of perihelion	167	31	33
Longitude of ascending node	72	50	50
Inclination	0	46	20
Tropical period (days)			30587.37
Mean distance (a.u.)			19.18254
Eccentricity			0.0461183
Mean daily tropical motion			42″.3704

In the summer of 1784 similar elements were computed by Johann Friedrich Wurm of Nürtingen (1760–1833). Baron Francis Xaver von Zach (1754–1832; court astronomer at Gotha), Fixlmillner, and others published tables of the planet's motion. Not long after, the special intractability of Uranus became rudely apparent.

By 1788, astronomers were reasonably confident that the representation of the Uranian orbit was past the difficult stage. Seven years of careful observation with modern instruments, and the positions recorded earlier by Flamsteed and Mayer, were thought to have provided all the data necessary to describe the path of Uranus with exemplary precision. The tables based on Fixlmillner's formulas might be subject to small corrections, but these could be regarded as refinements of a basically correct theory. It was a distinct surprise when, in 1788, the tables failed to represent the observed positions of Uranus with even tolerable accuracy.

Fixlmillner tried to resolve the difficulty by reconstructing his elements and tables, and in the process he encountered for the first time a dilemma that was to thwart astronomers for the next fifty years. He found that for any extended period his theory could satisfy either the old observations of Uranus or the new—but not both. He finally decided on an expedient solution. Since he could not reconcile Flamsteed's observation of 1690 with the positions recorded since 1781, Fixlmillner discarded the early observation and constructed new tables without it.

Other theoreticians came to the rescue. The Italian astronomer-priest Barnaba Oriani (1752–1832) had already worked on the Uranian orbit. After the discrepancy of 1788 Oriani tried to reincorporate all the existing observations of Uranus, including Flamsteed's, into the theory by calculating the perturbation effects of Saturn and Jupiter. The same project was taken up by Franz Joseph von Gerstner (1756–1832), and in the politically taut Paris of 1789 Lalande and Jean Baptiste Joseph Delambre (1749–1822) joined the mathematical assault on the wayward orbit.

The reform was successful. Delambre's tables, crowned by the French Academy in 1790, satisfied all modern observations and the two old positions recorded by Mayer and Flamsteed as well. These, however, were no longer the only "ancient observations"; in 1788, Pierre Charles Lemonnier (1715–1799) found that he had observed Uranus as a star in 1764 and 1769, and his additions did not prove unique. In the following thirty years sixteen more preplanetary observations of Uranus were discovered.

URANUS, 1781–1809

During this period the new planet was observed fewer and fewer times each year. Although the political uproar in Europe was responsible for some of the attrition, work on Uranus fell off in far greater proportion than that on other objects. At Greenwich, for instance, where a full program of telescopic astronomy was carried on without interruption, observations of Uranus steadily decreased from the dozens made in the four years following its discovery to only one in 1798. From then until 1814 one or two were made at each opposition, and there was only a trifling increase during the following fifteen years. Uranus was clearly a less novel object in the 1790's than in the decade of its discovery and Delambre's apparently correct tables allowed astronomers a welcome (but, in this case, temporary) respite. Toward the end of the eighteenth century attention was also distracted from Uranus by another insistent problem in planetary mechanics: the gap in Bode's series.

Ever since the appearance of Kepler's *Mysterium Cosmographicum* in 1596 the search for a simple mathematical series relating the orbital radii of the planets had been a preoccupation of German astronomers. In 1741 the physicist and mathematician Christian Freiherr von Wolf (1679–1754) pointed out that, allowing some exceptions, a sort of progression did exist. His ideas were taken up by his young disciple, Immanuel Kant (1724–1804), and were extended by Johann Daniel Titius (1729–1796; "Titius of Wittenberg"). In a prefatory note to his German translation of *Contemplation de la Nature* (edition of 1772), by the Swiss naturalist Charles Bonnet (1720–1793), Titius stated that, taking the Earth's distance from the Sun as 10, the mean radial distances of the planets were almost exactly proportional to the terms of the expression $4 + 3(2^n)$, except that there was an adjusted first term and no known planet corresponding to $n = 3$. The ratios were as follows: [8]

Planet	*Observed*	*Titius*
Mercury	3.8	4
Venus	7.2	7
Earth	10.0	10
Mars	15.2	16
—	—	28
Jupiter	52.0	52
Saturn	95.5	100

In 1772 Johann Bode came across this progression, enthusiastically adopted it, and incorporated it in the second edition of his introductory astronomy book.[9] Convinced of its validity, Bode soon became an astronomical evangelist for the formula that was eventually named after him. In a retrospective note, he later remarked, "In all my . . . astronomical writings I have, when occasion arose, spoken of this progression, presented it in sketches, and advanced many arguments for its correctness." [10] Bode was especially struck by the discontinuity in the series at $n = 3$ and the incidental correspondence of a seemingly disproportionate space between the fourth and fifth planets. He was not the first to notice this. Johann Kepler, in his quest for a numerically ordered solar system, had actually postulated the existence of an unobserved planet to fill the gap separating Mars and Jupiter. Bode, perhaps serving the same urge, advanced the same hypothesis in 1772.

By this time, many astronomers felt that coincidence alone was not enough reason to accept a hypothesis. Since his adopted law appeared to be only an interesting coincidence, devoid of any causal relation, Bode later had to write in a tone that was half chagrin, half chauvinism, "It is noteworthy that as yet no mention has ever appeared of this progression in the astronomical work of foreigners. Only German astronomers have mentioned it after I drew attention to it in my astronomical writings."[11] Not many astronomers were inclined to believe in the existence of Bode's unseen planet, invented to fill a term in his series, until they had more confidence in the series itself. It was William Herschel who unwittingly tipped the balance of opinion in Bode's favor.

Herschel's discovery of Uranus had come as a complete surprise; it was certain that no astronomer had any advance knowledge of the new planet before 1781. We have seen that by 1783, Fixlmillner, using observational data, had calculated the mean orbital radius of Uranus as approximately 19.2 astronomical units. In the Bode-Titius system, where the astronomical unit was multiplied by 10, Fixlmillner's result would be expressed as 10×19.2 or 192. Now Bode's law ostensibly offered a very simple method of predicting the orbital radius of any planet found beyond Saturn. One had only to solve the expression $4 + 3(2^n)$, substituting the appropriate power for n. The exponent n for Saturn was 5; if $n = 6$, for the next farther planet, was substituted, the formula gave $4 + 3(64)$ or 196 for the orbital radius of Uranus. Thus the radius predicted by Bode's law agreed within 2 percent with the radius calculated from observations. The prophet was raised up; a long era of scientific respectability lay ahead for the empirical black sheep of astronomical formulas.

The apparent confirmation of Bode's law reinforced his belief in the existence of an undiscovered trans-Martian planet. To his unshaken faith was lent the newly strengthened convictions of von Zach, who in 1785 actually tried to compute what he termed "analogical" elements for a planet accessible neither by sight nor by inferential effects. Von

Zach believed that the missing planet could be found by a methodical telescopic search of the small stars of the zodiac, and in 1787 he began such a search. He soon decided that the project was impossible for a single observer to carry out. Rather than abandon the idea he kept it intermittently before him for thirteen years, and finally he found what seemed to be a workable plan.

During a visit to Lilienthal in the autumn of 1800 von Zach met with five other German astronomers (Schröter, Harding, Olbers, and probably von Ende and Gildemeister) and founded the "Lilienthal Detectives," a society whose sole purpose was to search for the hypothetical planet. The group decided to divide the zodiac into twenty-four equal zones and to assign the search of each zone to a different European astronomer. Detailed star maps of each sector had just been completed and were being distributed to the various observers when one of the twenty-four, working alone and as yet uninformed of his part in von Zach's project, found the missing planet.

Giuseppe Piazzi (1746–1826), like William Herschel before him, was not looking for planets when he made his discovery. Working at Palermo, Sicily, Piazzi was observing stars for the catalogue he had set out to compile nine years earlier. On the night of January 1, 1801, he recorded the position of an eighth-magnitude star in the constellation Taurus. Piazzi customarily observed the same fifty stars on four consecutive nights; the following evening it seemed to him that the small star in Taurus had shifted slightly to the west. The next night he was certain it had moved. He measured its daily motion as 4^{m} in right ascension and $3' 30''$ in declination and began to observe it regularly. On January 14, the object went from retrograde to direct motion and Piazzi tracked it until February 11, when he was forced to stop because of severe illness.

On January 24, Piazzi had sent letters describing his discovery to Oriani in Milan, to Lalande in Paris, and to Bode in Berlin. The letter to Bode, which arrived on March 20, provided a classic example of a historically persistent astronomer's foible and a dubious physical principle: The Conservation of Least Planets. Despite conclusive evidence, Piazzi, like many of his colleagues, was exceedingly reluctant to acknowledge a new member of the solar system. He wrote to Bode (twenty years after the analogous discovery of Uranus), "On the 1st of January I discovered a comet in Taurus in right ascension 51° 47′, northern declination 16° 8′ . . . It is very small, and equivalent to a star of the eighth magnitude, without any noticeable nebulosity." [12] To Lalande he reported that he had discovered a very small comet without tail or envelope (!). Only to his close friend Barnaba Oriani did Piazzi confide that he supposed he had found a planet.

Not unexpectedly, von Zach was overjoyed with Piazzi's news. He published a detailed article entitled "On a long supposed, now probably discovered, new major planet of our solar system between Mars and Jupiter," [13] despite the fact that the planet had not yet been observed by anyone but its discoverer. By the time the astronomers of northern Europe heard of it, even Piazzi himself could no longer see the object. It had approached too near the Sun, and, as closely as could be predicted, would not be visible again until September of 1801.

Although the members of the Lilienthal group enthusiastically accepted the new planet, other astronomers were less excited, and at least one scholar was decidedly antagonistic. The postal delay between Palermo and Berlin allowed just enough time for a young philosopher named Georg Wilhelm Friedrich Hegel (1770–1831) to publish at Jena his *Dissertatio philosophica de orbitis planetarum,* in which he proved by the clearest logic that the number of planets could not exceed seven and exposed the misguided thinking of certain astronomers who induced the existence of a new planet merely to supply a missing term in a numerical series. Narrow fields of view were not restricted to those who looked through telescopes.

During the summer of 1801 Bode and others realized that the new discovery was lost. The figures that Piazzi had sent to Berlin were inadequate to define the planet's motion. After an involved exchange of notes a complete record of the Palermo observations was published by von Zach in his *Monatliche Correspondenz* for September 1801. Meanwhile, nearly every prominent German astronomer tried to predict the missing planet's position by calculating elements based on the meager available data. Von Zach labored over an approximately circular orbit; Johann Karl Burckhardt (1773–1825), after failing to represent the observations by a parabola, computed an ellipse that proved equally unsatisfactory, and Heinrich Wilhelm Matthias Olbers (1758–1840) attempted the same calculation with no success. Observation proved just as disheartening as theory. From the middle of August 1801 until the end of September observers all over Europe tried in vain to find the missing planet as it moved out of the Sun's glare. They struggled against one obvious handicap—the unusually bad weather—but they were unaware of an even more significant one: the incorrect orbital elements thus far computed were directing the search to the wrong part of the sky.

The true orbit of Piazzi's planet was finally calculated by a young theorist from Brunswick; his solution heralded a major advance in the techniques of mathematical astronomy. Karl Friedrich Gauss (1777–1855) had by his twenty-third year established himself as a brilliant mathematician.

Although deeply involved with researches in pure mathematics and lunar theory, Gauss was intrigued by the account of Piazzi's discovery in the September 1801 *Monatliche Correspondenz.* He temporarily put aside his other work and attempted to compute a suitable orbit. During September and October he concentrated on method. His first calculations specifically for the new planet were made early in November 1801, and later that month he sent his final elements to von Zach.

Gauss did not merely calculate a single elliptical orbit; in two months he developed completely new, radically efficient methods for computing all orbits. Some idea of his advance can be gained by considering that the computation for the orbit of a comet which had taken Euler three days of continuous work took Gauss one hour. Like Newton, Gauss made no hypotheses, but tried only to find the orbit that fitted the observations as closely as possible. To accomplish this, he had to solve the general problem of an ellipse free of all arbitrary assumptions. He succeeded in 1801, but his solution was only the beginning of a revolution that he was to complete eight years later.

The practical results of Gauss's work were apparent almost immediately. Von Zach, armed with the Brunswick mathematician's ephemeris for the missing planet, now directed his persistent observations to the constellation Virgo. He caught a glimpse of a suitable object on December 7, but the weather prevented him from verifying his suspicions. On the last night of 1801, a bitter frost set in, the sky cleared, and there in Virgo, almost exactly in the position Gauss had predicted for the missing planet, von Zach found a strange star. The following night—the anniversary of the discovery—von Zach verified that Piazzi's lost planet had been found. Olbers rediscovered it independently that same night.

The astronomers who had believed in the existence of the new planet were vindicated; Piazzi was delighted. For his discovery he chose the name Ceres Ferdinandea, after his ruler and the titular goddess of Sicily.[14] Gauss calculated the mean distance of Ceres as 2.767 astronomical units, or, in Bode's system, 27.67; the term for the missing planet in Bode's series was 28. The empirical law was confirmed again, and the Lilienthal group exulted that the unseen planet first postulated by Kepler had been found at last.

It was the word "major" that caused the first doubts about Ceres. "Haupt" appeared prominently in the title of nearly every German publication about the new planet, but when William Herschel attempted to measure Ceres's diameter later in 1802 he arrived at the surprising figure of only 161.6 miles. Obviously a body of such minute size could hardly be called a major planet. (During his tests, Herschel was so disconcerted by the tiny size of Ceres that he doubted his telescope's magnification. As a check, he attempted to compare the size of Ceres with that of Jupiter by projection, and found that at a distance where the image of Ceres would have measured 0.71 inch across, the apparent diameter of Jupiter was 12 feet 11 inches.) A second anomaly was revealed by Gauss's elements for Ceres. Its orbit was inclined to the ecliptic at an angle of more than 10°, exceeding that of any other planet. This disclosure was overshadowed by an even more surprising announcement which deprived Piazzi's discovery of its unique position altogether.

During his search for the truant planet, Olbers had acquired an intimate knowledge of the smaller stars in the constellation Virgo. On March 28, 1802, while observing this part of the zodiac, he noticed a seventh-magnitude star that seemed unfamiliar. Tracking it carefully, he found that he had joined the august company of Herschel and Piazzi: the star moved. Olbers took a week to confirm his discovery, then sent his observational data to von Zach in Seeberg and to Gauss in Brunswick, requesting that the latter compute an orbit for the new planet. When Gauss finished the problem, on April 18, his elements included a startling figure for the mean orbital radius: 2.670 astronomical units. Apparently, Olbers's discovery, which he named Pallas, was revolving at nearly the same distance from the sun as was Ceres. Gauss found that the orbit of Pallas was remarkable in other ways. Its eccentricity was 0.24764, even exceeding that of Mercury, and its inclination to the ecliptic was an unprecedented 34° 39′. Herschel's measurement of diameter, which followed shortly, made Pallas only 110.3 miles across.

There were no historical precedents for this perplexing situation. Instead of the expected major planet, astronomers had found two small objects moving in unusual orbits between Mars and Jupiter. Herschel, for one, could not accept it. In a systematic, if biased, comparison between the characteristics of stars, planets, comets, and the newly discovered objects, he rejected all the traditional classifications for celestial bodies and proposed the title "asteroids" for Ceres and Pallas. (This misleading name has been supplanted by "planetoid" or "minor planet" in recent years.) Conservatism and historical influence played a major part in this proposal; Herschel showed how far the fortunes of Bode's law had risen when he wrote, in 1802:

> There is a certain regularity in the arrangement of planetary orbits, which has been pointed out by a very intelligent astronomer, so long ago as the year 1772; but this, by the admission of the two new stars into the order of planets, would be completely overturned; whereas, if they are of a different species, it may still remain established.[15]

Olbers, faced with the same difficult choice, found another solution. He knew that the observed properties of Ceres and Pallas were those of planets, however small, and he believed in Bode's law. He tried to reconcile the apparent conflict with a radical and ingenious hypothesis. Olbers suggested that Ceres and Pallas were small fragments (or in Agnes Clerke's inimitable phrase, "cosmical potsherds") of a large trans-Martian planet that had violently disintegrated at some time in the remote past. The explosion, which Olbers believed could have been caused by collision with a comet or by powerful internal forces, would have produced many small bodies moving in various orbits, but all with a common point of intersection at the scene of the original catastrophe.

The proof of Olbers's hypothesis was implicit in its statement. Obviously, if the primitive planet had been of any size, Ceres and Pallas would not be its only surviving fragments. (Herschel had calculated that it would require 31,000 bodies the size of Pallas to make up the volume of Mercury.) Olbers pointed out that the orbits of all other remnants of his postulated planet would have to pass through two opposite nodes, one located in Cetus and the other in Virgo, and suggested that monthly searches be made of these areas. He set the example himself, but despite two years of arduous work he found nothing.

Meanwhile, Karl Ludwig Harding (1765–1834), one of Schröter's assistants at the Lilienthal observatory, had begun to construct a series of charts showing all the small stars near the paths of Ceres and Pallas, in order to make the recognition of the minor planets less difficult. On September 2, 1804, Harding observed a faint star in the constellation Pisces, very close to that part of Cetus where Olbers had predicted the fragments of the primitive planet must pass. Two nights later the star had moved, and successive observations confirmed that it was indeed another planet. Harding named it Juno, and Gauss quickly computed its elements. He found that, like its two predecessors, it revolved between Mars and Jupiter with a mean orbital radius of 2.670 astronomical units, and that it had an extreme eccentricity of 0.2543.

Olbers was much encouraged by this confirmation of his prediction, and he continued his zodiacal observations with renewed confidence. His patience was ultimately rewarded on the night of March 28, 1807, when he discovered a fourth minor planet in Virgo, again as predicted. Gauss was given the honor of naming the new discovery, in recognition of his earlier labors on the orbits of the small planets, and he chose the name Vesta.

By this time Gauss had brought the methods originally developed for computing the orbit of Ceres to an extraordinary degree of refinement. In March 1809 they were published in the *Theoria motus corporum coelestium in sectionibus conicis solem ambientum.* Unwilling to release what he considered imperfect work, Gauss had taken eight years to produce a book that was immediately recognized as the beginning of a new epoch in mathematical astronomy and a scientific work of art. Gauss was well aware of his own perfectionism. In 1801 he had solved the general problem, "To determine the orbit of a heavenly body, without any hypothetical assumption, from observations not embracing a great period of time, and not allowing a selection with a view to the application of special methods." [16] His friends had urged him to publish the solution at once, but Gauss had demurred. In the preface to the *Theoria motus* he gave his reasons for having done so: "other occupations, the desire of treating the subject more fully at some subsequent period, and, especially, the hope that a further prosecution of this investigation would raise various parts of the solution to a greater degree of generality, simplicity, and elegance."

His hope was realized: in the *Theoria motus* astronomers found methods of unprecedented conciseness and elegance for the determination of an orbit from three observations; for the determination of an orbit from four observations, of which only two were complete; and—most general of all—for the "Determination of an Orbit satisfying as nearly as possible any number of Observations whatever." In addition, the book contained Gauss's presentation of the method of least squares, which found application in many fields besides astronomy. Stimulated by a single small planet, Gauss had evolved a complete new system for the analysis of scientific data.

CHAPTER 2

URANUS, 1809–1840

THE publication year of the *Theoria motus* marked the beginning of a hiatus in astronomical discovery. There was a distinct lessening of interest in the minor planets by 1809, for despite two years of meticulous observation after the discovery of Vesta no additional "fragments" of Olbers's primitive planet were found. The first group of four discoveries had taken only six years; new planets had followed one another with sensational rapidity. Now, after what seemed a long unproductive lapse, there was a tendency to regard the possibilities as exhausted. Although Olbers expressed his confidence that in time other planetoids would be found, William Herschel's son John (1792–1871) passed pompous judgment on the idea: "This may serve as a specimen of the dreams in which astronomers, like other speculators, occasionally and harmlessly indulge." [1] Eventually, on December 8, 1845, Herschel was to be proved wrong, but Olbers did not survive to witness his vindication.

Olbers's hypothesis of a primitive planet found an enthusiastic supporter in Lagrange, who in 1814 attempted to investigate the physics of a planetary explosion, and showed that the violent disintegration of a planet into small fragments was feasible in terms of terrestrial forces. He calculated that an initial speed of less than twenty times the muzzle velocity of a cannon ball (a somewhat arbitrary reference standard) would have put the individual particles into elliptic orbits.

Bode's law prospered even more than Olbers's hypothesis. Between 1800 and 1807 it advanced from an interesting but slightly disreputable speculation to an accepted rule on which future planetary theory was to be confidently based.

There was not much planetary theory completed in Europe during the second decade of the nineteenth century. The Napoleonic wars and the wars of liberation interrupted scientific work in nearly every country. The demands of patriotism and sometimes plain survival superseded abstract calculations. Gauss, who watched his mortally wounded ducal patron leave Brunswick after the battle of Jena, was typical of many scientists in that he found it a difficult time for contemplation. Not until after the second Peace of Paris in November 1815 did international tension ease to the point where scientific communication could be resumed and investigations pursued without fear of interruption.

Many astronomers used their libraries as a refuge from political chaos. Although few contemporary positions of Uranus were recorded between 1800 and 1820, the search for ancient observations was continued during the same period with remarkable success. In 1813 Friedrich Wilhelm Bessel (1784–1820), director of the Königsberg observatory, found, while reducing James Bradley's observations, that Bradley had observed Uranus as a star on December 3, 1753. Johann Burckhardt subsequently discovered that Flamsteed had recorded the position of Uranus not once, but three times. To his observation of 1690 there was now added one of April 2, 1712, and one of April 29, 1715. (Some years later Burckhardt found three more observations of Uranus made by Flamsteed on March 4, 5, and 10, 1715.) In 1820, Alexis Bouvard (1767–1843) searched the records of his deceased colleague, Pierre Charles Lemonnier (1715–1799) and found ten more observations of

Uranus that had been made before 1781. In January 1769 Lemonnier had observed Uranus six times in nine days, and if he had reexamined his notes he probably would have discovered the new planet. Lemonnier did not, however, keep very tidy notes; one of his observations of Uranus was recorded on a paper bag that had originally contained hair powder. It was a novel and thrifty system, but not one that facilitated revealing comparisons. Early in 1820 the known ancient observations of Uranus were as follows: in 1690, by Flamsteed, December 23; in 1712, by Flamsteed, April 2; in 1715, by Flamsteed, April 29; in 1750, by Lemonnier, October 14, December 3; in 1753, by Bradley, December 3; in 1756, by Mayer, September 25; in 1764, by Lemonnier, January 15; in 1768, by Lemonnier, December 27, 30; in 1769, by Lemonnier, January 15, 16, 20, 21, 22, 23; in 1771, by Lemonnier, December 18.

Bouvard's additions to this list were not, like Bessel's, a by-product of other researches. By this time it had become obvious that Delambre's tables of 1790 for Uranus were no longer in satisfactory agreement with observation. Bouvard had, in 1808, published tables for the motions of Saturn and Jupiter, but, owing to errors in the theory and computed mass of the two planets, his figures had proved unreliable. His researches of 1820 were part of an effort to correct both his own and Delambre's errors, and to calculate new tables for all three exterior planets that would represent their motions accurately. The revisions for Saturn and Jupiter were completed successfully, but when Bouvard attempted to correct the tables for Uranus he found himself in the same dilemma that had plagued Fixlmillner thirty years earlier.

Fixlmillner, it will be remembered, had great difficulty reconciling the observations of Uranus made between 1781 and 1788 with the two ancient observations (1690 and 1756) then known. Bouvard's problem was identical, only greatly magnified: instead of two ancient positions, he knew of the seventeen listed above. Where Fixlmillner had used seven years of modern observations, Bouvard's comparable data spanned almost four decades. Moreover, although Fixlmillner had abandoned the conflicting position of 1690, Delambre had afterward succeeded in temporarily reconciling all the extant observations. In 1820 this no longer seemed possible.

Bouvard introduced his new tables with the announcement that he had not been able to find any single elliptic orbit that would incorporate both the ancient and modern observations of Uranus:

> The construction of the tables of Uranus involves this alternative:—if we combine the ancient observations with the modern ones, the first will be adequately represented, but the second will not be described within their known precise tolerances;—while if we reject the ancient positions and retain only modern observations, the resulting tables will accurately represent the latter, but will not satisfy the old figures. We must choose between the two courses. I have adopted the second as combining the most probabilities in favor of truth, and I leave to the future the task of discovering whether the difficulty of reconciling the two systems results from the inaccuracy of the ancient observations, or whether it depends on some extraneous and unknown influence which may have acted on the planet.[2]

It was a difficult choice to make, and Bouvard did not have the conviction to let his decision stand on its own merits; he included in the introduction to his tables a shabby plea that he had rejected the ancient observations because they were inaccurate and wholly unreliable. He was, in effect, accusing four first-class observers, whose records were accepted as a matter of course, of coincidentally making extraordinary errors in determining the position of the same star. Although there was no proof whatever to support such a contention, it was not rejected as quickly as it should have been.

One reason for this was that by rejecting the ancient observations Bouvard unwittingly provided Delambre with a face-saving device for his own superseded tables. Delambre eagerly seized on the idea. He furthered its acceptance (and his own reputation) by writing in his widely read *Histoire de l'Astronomie:*

> We see that M. Bouvard, after working on the problem for a long time, was unable to reconcile everything, and finally rejected all the observations in which the planet had been taken for a fixed star. He did not even use Mayer's one observation, which I had included out of pure respect for that great astronomer, and which certainly made my own tables less accurate and less permanent.[3]

It is difficult to determine how many astronomers were convinced by Delambre's pious testimony that he had used Mayer's observation of Uranus "out of pure respect." The hypocritical compliment to Mayer was even less justified than Bouvard's slur on the accuracy of Flamsteed, Bradley, and Lemonnier.

There was little to recommend the latter action, either morally or scientifically. Bouvard's tables attributed errors in angular measurement of 65″.9, 60″, and 40″ to astronomers whose observations were generally (and correctly) held to be accurate within 10″, 6″, and 5″, respectively. Bouvard was thus accusing the "ancient" observers of enormous percentage errors; 659 percent, 1000 percent (that is, an order of magnitude), and 800 percent. No error approaching these levels had ever been found in the records of any of the men in question. Even with such questionable disqualifications, Bouvard's tables—for which the whole rationalization had been made—allowed a maximum deviation from the modern observations of 9″, or about twice the accepted contemporary accuracy.

Neither Friedrich Bessel nor the French astronomer Urbain Leverrier (1811–1877) was converted by Bouvard's arguments. As early as 1823 Bessel began to make new observations of Uranus, noting any differences between the observed positions and those specified by the French tables. He also reviewed Bouvard's calculations closely and found four theoretical errors, two of which altered every term in the particular tables dependent upon them. Bouvard acknowledged the mistakes and supplied the necessary corrections in a rather cool note to Bessel. The discrepancies involved were not large, but they confirmed some of the latter's suspicions. Bessel refuted Bouvard unsparingly in a public lecture given at Königsberg on February 28, 1840:

> In my opinion, Bouvard made much too light of the matter; inasmuch as, after he had found himself unable to reconcile the theory both with the ancient observations and the forty-year series of modern ones, he contented himself with the remark that the former were not so accurate as the latter. I have myself subjected them to a more careful investigation and new calculation, and have thereby attained the strong conviction that the existing differences, which in some cases exceed a whole minute of angle, are by no means attributable to the observations.[4]

Leverrier passed a similar judgment. Although the following passage was written after the fact, when it was known that the errors in the ancient observations did not exceed 9″ (exactly the same maximum error that Bouvard's tables allowed), Leverrier's actions during the intervening time left no doubt that he had held the same opinion all along:

> One should note that, though the observations of Flamsteed, Bradley, Mayer, and Lemonnier are not as exact as those of

the astronomers of our epoch, one may not with any plausibility be allowed to consider them infested with such enormous errors as those of which the present tables accuse them. The author of these tables [Bouvard] actually suggests, however, that this is his opinion.[5]

Those who were skeptical of Bouvard's assumptions did not have long to wait before their suspicions were verified. Less than four years after the new tables were published, significant discrepancies began to appear between Bouvard's figures and the observed positions of Uranus.

The first corrections were reported from the observatory at Kremsmünster by Bonifacius Schwarzenbrunner (1790–1830). Following the opposition of 1825, which Schwarzenbrunner fixed at $4^h\ 14^m\ 0^s$ on July 10, Uranus was observed at Kremsmünster for two weeks. On July 15, significant differences from Bouvard's figures were recorded: in declination, 13″.8; in geocentric latitude, 15″.0. The deviations reached a peak on July 20 and 23, when right ascensions of Uranus were recorded that differed from the tables by $17^s.8$ and $17^s.3$, respectively. Although these were unusually large discrepancies, they represented ony four of forty-two observations recorded; of the remaining thirty-eight positions, twenty-seven agreed with the tables within 5″, which may be taken as a reasonable standard of contemporary accuracy. The average corrections of seven observations for each of six coordinates were: −3″.5, −6″.7, −2″.4, −7″.7, −2″.4, −7″.3. These values were fairly satisfactory and caused little apprehension.

By the following year the disagreement seemed worse. In 1826, according to Schwarzenbrunner, the opposition occurred on July 14, at $18^h\ 15^m\ 42^s$, and observations were made at Kremsmünster from July 6 through July 20. Of the fifty-four positions recorded during this time, only twenty-three were within the 5″ tolerance of agreement with the tables. The six average deviations, of nine observations each, had now risen to: −6″.5, −10″.1, −4″.2, −10″.9, −4″.0, −10″.4. Such figures were definitely out of line and provoked comment by many astronomers, both on the continent and in England.

There would probably have been more pressure for revision at this time if the situation had not seemed to improve. The true (heliocentric) longitude of Uranus, which had remained steadily in advance of the computed longitude, suddenly began to drop at an increasingly rapid rate until, in 1829–30, the tabular and observed longitudes coincided. This singular agreement distracted astronomers from the unexplained errors already found in the tables and from the peculiar variations of the differences, but any reassurance provided by the coincidence of the figures for true longitude was short lived. Far from maintaining the much-hoped-for agreement, the planet began to fall behind its calculated place more and more rapidly. The differences soon increased to the point where the tables could no longer be accepted as representing even approximately the true motion of Uranus. In 1832, the English astronomer George Biddell Airy (1801–1892) reported to the British Association for the Advancement of Science that Bouvard's tables, constructed only eleven years earlier, were currently in error (in geocentric longitude) by nearly a half-minute of arc. Since Uranus was the only celestial body subject to such a large discrepancy, the pressure to devise some explanation was overwhelming.

Five different hypotheses for the strange behavior of the seventh planet were prevalent in the early 1830's, but by the end of the decade three of these had been discarded, and the fourth was rapidly losing adherents. The first three suggestions either were far-fetched or were expedient reappearances of periodically fashionable theories. One of these was the Cartesian hypothesis of a cosmic fluid. Its proponents, as before, ignored the fact that its existence had never been proved and that, even if it had, it could not account by any known laws for the observed perturbations of Uranus. A second, less arbitrary, idea was the supposed existence of an undiscovered massive satellite of Uranus. It was soon pointed out that such a satellite would cause perturbations of much shorter period than those shown by Uranus, and, allowing the known range of satellite densities, would be large enough to be visible through a telescope. The third proposal was advanced to save the tables—rather than the phenomena—and retrospectively appears to be something of a desperation measure. This was a "catastrophe" hypothesis, in which it was assumed that a comet had struck Uranus very close to the time of its discovery in 1781, sensibly changing its orbit. Thus the planet's movements would be correctly represented by one ellipse prior to 1781 and by a different one subsequent to that year, as Bouvard had stated. The steadily increasing errors of the tables based on the "second" ellipse sufficed to discredit this idea.

A more difficult question was raised by the proponents of the fourth hypothesis, who suggested that the law of gravitation, upon which Bouvard's tables were indisputably based, might differ in some way at the tremendous radial distance of Uranus from the Sun. There was plenty of precedent for such a doubt; since the beginning of the eighteenth century, alteration of the Newtonian law had provided a dependable escape hatch for theoreticians trapped by their own hypotheses.

An outstanding instance of such doubt had occurred in 1748, when the French Academy of Sciences offered a prize for an investigation of the perturbations of Saturn and Jupiter. The analytical triumvirate of Euler, Clairaut, and d'Alembert submitted competitive manuscripts on the three-body problem. In the work on lunar theory that developed out of this competition, each of the contestants arrived independently at the conclusion (reached previously by Newton) that the theoretical motion of the lunar perigee was only half that determined from observation. Unable to account for the discrepancy, Clairaut proposed a new law of gravitation which combined an inverse-square term with a term for the inverse-fourth power of the distance. The great French naturalist Georges Buffon (1707–1788) argued against such a law, but only when Clairaut discovered that he had neglected certain small but significant factors did he abandon his expedient construct and recompute the formula. Both he and Euler were convinced that the revised calculation agreed perfectly with Newton's law and with observation; the apparent flaw in the law of gravitation had become a brilliant testament to its correctness. A second, nearly identical, astronomical fairy tale was played out in 1770–1787. The Moon's secular acceleration was the problem, Euler was the doubting villain, and Laplace was the analytical hero. Once more Newton's law was triumphantly confirmed.

The unexplained motions of Uranus raised new doubts about the validity of the law of gravitation. The idea of a slightly diminished law of force at extreme distances from the Sun appealed to some astronomers, among them George Biddell Airy. Airy, who was appointed Astronomer Royal on August 11, 1835, wrote to the Reverend Richard Sheepshanks (1794–1855), a fellow member of the Royal Society, on December 17, 1846, that there was "great probability" that "the law of force differed slightly from that of the inverse square of distance." [6] Even in the 1830's, only a very few astronomers believed in such a probability; Friedrich Bessel was typical of the distinguished contemporary adherents to the Newtonian system. Neither Leverrier nor John Couch Adams (1819–1892), both of whom were to do outstanding work on the perturbations of Uranus, had any doubts about the validity of the law of gravitation. The strong historical case for its correctness eventually won out, and the weight of informed opinion concentrated on a fifth hypothesis to explain the strange behavior of Uranus: the existence of an undiscovered exterior planet.

This was not a new idea. Perhaps the first public statement of such a possibility was made by Clairaut in November 1758. Appended to his prediction for the return date of Halley's comet was the remark "that a body which travels into regions so remote, and is invisible for such long periods, might be subject to totally unknown forces, such as the action of other comets, or even of some planet too far distant from the Sun ever to be perceived." [7] Clairaut's remark was appropriately received as pure speculation, but it was a provocative and open-minded conjecture. The possible existence of an eighth planet was not seriously considered until the unexplained discrepancies of Uranus became too large to ignore. As we have seen, in 1821 Bouvard had mentioned "extraneous and unknown influences" acting on Uranus. He later corresponded with Peter Andreas Hansen (1795–1874), director of the Seeberg observatory, on the subject of an exterior planet, and was definitely converted to that hypothesis some time before 1836.

After Airy's unsettling report of 1832, one of the first documented commitments to the idea of a trans-Uranian planet was made by a British amateur astronomer, the Reverend Dr. Thomas John Hussey. On November 17, 1834, Hussey wrote to Airy:

With M. Alexis Bouvard I had some conversation upon a subject I had often meditated, which will probably interest you, and your opinion will determine mine. Having taken great pains last year with some observations of Uranus, I was led to examine closely Bouvard's tables of that planet. The apparently inexplicable discrepancies between the ancient and modern observations suggested to me the possibility of some disturbing body beyond Uranus, not taken into account because unknown. My first idea was to ascertain some approximate place of this supposed body empirically, and then with my large reflector set to work to examine all the minute stars thereabouts: but I found myself totally inadequate to the former part of the task . . . I therefore relinquished the matter altogether; but subsequently, in conversation with Bouvard, I inquired if the above might not be the case: his answer was, that, as might have been expected, it had occurred to him, and some correspondence had taken place between Hansen and himself respecting it. Hansen's opinion was, that one disturbing body would not satisfy the phenomena; but that he conjectured there were two planets beyond Uranus. [Hansen later denied ever having made such a statement.[8]] Upon my speaking of obtaining the places empirically, and then sweeping closely for the bodies, he fully acquiesced in the propriety of it, intimating that the previous calculations would be more laborious than difficult; that if he had leisure he would undertake them and transmit the results to me, as the basis of a very close and accurate sweep.[9]

As we shall see, Hussey was discouraged from making such a sweep by Airy's pessimistic reply, but in the following year other astronomers suggested the existence of an eighth planet. Halley's comet, which had provoked Clairaut's speculation seventy-seven years earlier, prompted Jean Élix Benjamin Valz (1787–1861) to make a more definite statement in 1835. After making careful observations of the comet, Valz decided that there were extraneous forces acting on it. Rejecting known sources of disturbance, he announced:

I would prefer to have recourse to an invisible planet, located beyond Uranus; its orbit, according to the progression of planetary distances [Bode's law], would be at least triple that of the comet, so that in a few cycles, the perturbations would be about the same, and the calculations of four or five intervals might enable us to recognize it. Would it not be admirable thus to ascertain the existence of a body which we cannot even observe? [10]

Friedrich Bernhard Gottfried Nicolai (1793–1846), director of the Mannheim Observatory, asked almost the same question as had Valz, as a result of the same stimulus. Nicolai's observations of Halley's comet were made between August and November 1835. He found that the time of perihelion occurred an entire day later than expected, and he noted that similar discrepancies had been recorded on previous visits. In the report of his observations, Nicolai wrote, "One immediately suspects that a trans-Uranian planet (at a radial distance of 38 astronomical units, according to the well-known rule) might be responsible for this phenomenon." [11]

The existence of such a body beyond Uranus was also suspected by Niccolo Cacciatore (1780–1841), director of the Palermo Observatory, at about the same time. In May 1835 Cacciatore found a moving star near Number 17, Hora XII of Piazzi's catalogue. Hampered by miserable weather, he was able to observe the star only three nights before it disappeared into the twilight. He estimated its movement in right ascension during that time as 10 seconds of arc. "Such a slow motion," he wrote, "makes me suspect that the [moving] star is located beyond Uranus." [12] Cacciatore's description of the star attracted little attention in the *Monthly Notices* of the Royal Astronomical Society, but a similar account published in the *Comptes rendus* of the French Academy provoked some interesting comments. After reading Cacciatore's letter, which appeared in the *Comptes rendus* on February 15, 1836, Louis François

Wartmann (1793–1864) wrote from his private observatory at Geneva that he too had seen a moving star, on September 6, 1831. Wartmann believed that he had observed a trans-Uranian planet, and submitted figures for its mean distance and period:

> It would seem most probable that this minute point is a planet, which travels around the Sun in an orbit of considerable radius . . . The new planet must needs lie at about double the distance of Uranus from the Sun, i.e., at a radius of 388, that of the Earth being 10; its period of revolution must therefore be about 243 years.[13]

Cacciatore's announcement elicited criticism from Olbers and Valz, both of whom felt that the Palermo observations did not provide enough evidence to support Cacciatore's belief that he had seen a planet. In Valz's opinion, it was "quite risky" to judge from motion alone. Olbers proposed some tests to determine what the object actually was, but he also wrote that, if it were a trans-Uranian planet, "it would confirm an old conjecture made by Mr. Bouvard and a few other astronomers." [14] The conjecture, as we have seen, was not really very "old." Olber's choice of adjective confirmed a changing climate of opinion.

The shift of intellectual loyalties was a protective and self-eradicating process; outdated allegiances were quietly jettisoned, and often they were carefully buried. By 1836 the trend of planetary astronomy was obvious, and men who had disparaged the exterior-planet hypothesis a few years earlier were describing it in sentences beginning "Of course." The works of the redoubtable Mrs. Somerville provided a splendid example of the change in scientific attitude toward a supposed eighth planet.

Mary Fairfax Somerville (1780–1872) was a remarkable and ambitious woman. As a young girl she learned Latin solely in order to be able to read Newton's *Principia.* She later mastered calculus and celestial mechanics, and manifested in general what her contemporaries regarded as "a masculine breadth of mind." Her second husband (and first cousin), Dr. William Somerville, brought her to London in 1816, and she rapidly became something of a scientific celebrity. George Airy and both John and William Herschel were among her good friends, and she corresponded with Laplace and with François Arago (1786–1853), director of the Paris observatory. Such friendships kept her in close touch with contemporary developments in astronomy. In 1830 Lord Brougham asked her to write a popular version of Laplace's *Mécanique céleste,* and she obliged with *The Mechanism of the Heavens.* The book was a tremendous success, and the author followed it with *The Connexion of the Physical Sciences,* which went through seven editions. It was in these books that Mary Somerville, gifted, and above all flexible, provided a record of the changing planetary theories of the 1830's.

The Mechanism of the Heavens was published in 1831; in that year the author knew exactly where she stood with regard to the planetary system. It was securely buttoned up, and Uranus marked its farthest boundary, for, she said, "If we judge of the distance of the planet by the slowness of its motion, it must be on the very confines of the Solar System." [15] She went on to explain how Bouvard's tables described the orbit of Uranus, and firmly assured her readers that "the only sensible perturbations in the motion of this planet arise from the actions of Jupiter and Saturn."

There were long chapters on perturbations and planetary tables in the first edition of *The Connexion of the Physical Sciences,* which appeared in 1834. By this time, Mrs. Somerville's conviction about Uranus had vanished altogether. The new book included a description of Herschel's discovery, but the section on planetary tables was badly out at the seams. It skipped from the tables of Jupiter and Saturn directly to a weak closing remark about the commendable accuracy of the tables for the minor planets, considering that they had been observed for so short a time. There was no mention of Uranus (or of Bouvard) at all. It was important for Mrs. Somerville to keep a neat conceptual house, and in 1834 Uranus was a decidedly untidy problem.

Two years later, most astronomers had accepted the hypothesis of an exterior planet, and Mrs. Somerville was solidly on the side of the majority. In the third edition of *The Connexion of the Physical Sciences* Uranus was restored to its rightful position, one that the author had taken care to furnish with several foolproof exits. In 1836 the ever-adaptable Mrs. Somerville wrote:

> The tables of Jupiter and Saturn agree almost perfectly with modern observations; those of Uranus, however, are already defective, probably because the discovery of the planet in 1781 is too recent to admit of much precision in the determination of its motions, or . . . possibly it may be subject to disturbances from some unseen planet revolving about the sun beyond the present boundaries of our system. If, after a lapse of years, the tables formed from a combination of numerous observations should still be inadequate to represent the motions of Uranus, the discrepancies may reveal the existence, nay, even the mass and orbit, of a body placed for ever beyond the sphere of vision.[16]

The complete rearrangement of Mrs. Somerville's mental furniture had taken only six years. After the first discrepancies of Uranus were reported, it took Alexis Bouvard almost twice that long before he decided, in 1837, to revise his tables. He began work on the elements of Jupiter and Saturn himself, and delegated to his young nephew, Eugène Bouvard, the reconstruction of the tables of Uranus. On October 6, 1837, Eugène wrote to George Airy about the reduction of new observations for the tables, and in the letter he confirmed his uncle Alexis's belief in an exterior planet. Referring to the difference between the tabular and the observed position of Uranus, he asked, "Can this be attributed to an unknown perturbation exerted on the motions of this star by a body situated beyond it? I don't know, but this at least is my uncle's notion." [17]

While Eugène Bouvard worked on the new tables, Airy continued with the reduction of planetary observations that was his main project in 1837. On February 24, 1838, he reported to Heinrich Christian Schumacher (1780–1850), editor of the *Astronomische Nachrichten,* that, compared with observations made at Cambridge in 1833, 1834, and 1835, and at Greenwich in 1836, the tabular radius vector of Uranus was "considerably too small." Airy was later to attach exaggerated importance to this fact.

After 1838 the exterior-planet hypothesis became public property, rather than the privately urged conviction of a few individuals. A contemporary American astronomer later wrote:

> The problem became, from this time forth, one of the most important questions of Physical Astronomy. Astronomers in various countries busied themselves with it and spoke of it without reserve . . . Numerous mathematicians subsequently conceived the purpose of entering earnestly into laborious and precise calculations, in order to determine whether the assumption of an exterior cause of disturbance were absolutely necessary, and, if so, to determine from the known perturbations their unknown cause.[18]

There were still some obstinate heretics, as well as some outstanding missionaries. One of the latter was Johann Heinrich von Mädler (1794–1874), director of the Dorpat observatory. In 1841 Mädler published his *Populäre Astronomie,* a book that would not be considered "popular" in the remotest sense by twentieth-century publishers. In this work Mädler advanced a well-buttressed argument for the existence of an eighth planet. He included a revealing analysis of the periodic differences between the calculated and the observed motions of Uranus, and pointed out that the current error of its radius vector was larger than the distance from the Earth to the Moon. Summing up his evidence, Mädler concluded,

> we arrive at a planet acting upon and disturbing [Uranus]; we may even express the hope that analysis will at some future time realize in this her highest triumph, a discovery made with the mind's eye, in regions where sight itself was unable to penetrate.[19]

Echoing this hope, Friedrich Bessel continued to support the hypothesis of which he had been such an early and staunch advocate. In the Königsberg lecture of 1840, he announced that in his opinion the deviations of Uranus were

> discordances whose explanation can only be found in a new physical discovery . . . Further attempts to account for them must be based on the endeavor to discover an orbit and a mass for some unknown planet, of such a nature that the resulting perturbations of Uranus may reconcile the present lack of harmony in the observations.[20]

Two years later, while visiting England, Bessel informed John Herschel that he was about to undertake the necessary investigation himself. Bessel planned to use new reductions of the observations of Uranus made by his pupil Friedrich Wilhelm Flemming, but Flemming died suddenly, and shortly thereafter Bessel himself was attacked by an illness that incapacitated him until his death in 1846.

In 1842, the Royal Academy of Sciences at Göttingen proposed as a prize question a full discussion of the motion of Uranus, with special attention to the large and increasing errors of Bouvard's tables. The problem was important to astronomers everywhere, but it was to have special significance for two young mathematicians, one in England and the other in France. The latter had, by 1842, already distinguished himself by his outstanding work on perturbation theory. His name was Urbain Jean Joseph Leverrier.

CHAPTER 3

• • • • • • • • • • • • •

LEVERRIER

URBAIN Jean Joseph Leverrier was born on March 11, 1811, in the town of Saint-Lô, Normandy. His father, a modestly salaried employee of the State Property Administration, believed strongly in the value of higher education and was extremely ambitious for his son. Leverrier was educated in the schools of Saint-Lô, and he completed the course of literary studies that was offered at the local college. Since he showed some aptitude for science, his parents sent him, when he was seventeen, to college at Caen for two years of mathematics. In 1830 Leverrier graduated at the head of his class and, encouraged by this success, he entered the difficult competition for entrance to the École Polytechnique in Paris. He was not admitted; his failure was a sharp disappointment to his teachers and his family. The setback stimulated the elder Leverrier to greater self-sacrifice. By selling his house, he managed to send his son to the Collège de Saint Louis, "to perfect at Paris an already very well-grounded education." During his first year in Paris, Urbain won his college's *Prix de mathématiques spéciales,* and in the *concours* of 1831 for the École Polytechnique he justified his father's faith by carrying off one of the highest honors.

Leverrier was never again to descend from the first rank of scholars. He worked very hard at the École Polytechnique; the impression he left on his classmates was that of a serious and industrious student, one with more tenacity of purpose than natural inclination for science. During his college years he showed no special preference for any particular field of scientific inquiry. A school contemporary later commented, "If a penetrating, solid, occasionally exceptional intellect, always ready for controversy gave certain promise of an honorable career, no one could then have predicted [Leverrier's] future brilliance." [1]

The government of France was reorganized during the time that Leverrier attended school in Paris. He came to the capital just after the revolution that ushered in the July Monarchy, and his first job resulted from Louis Philippe's policy of employing graduates of the École Polytechnique in the government. Since Leverrier graduated with great distinction, he was rewarded with a choice of working for any one of several departments. A budding interest in chemistry and the lure of a great name, Joseph Louis Gay-Lussac (1778–1850), induced Leverrier to choose the Administration of Tobacco.

Gay-Lussac's laboratories were located at l'École d'application du Quai d'Orsay. Leverrier worked there for three years and showed great promise as an experimental chemist. His first paper, on the combination of phosphorus with hydrogen, was published in 1835, and a second article, on the reactions of phosphorus with oxygen, appeared two years later.[2] By that time, the tone of Leverrier's writing was no longer that of a beginner; he had become a mature scientist.

He had also become something of a sophisticate. Life in the capital was very much to Leverrier's taste, but his fascination with Paris forced him into making a difficult decision. As a civil servant he was required to do field work in various outlying regions of France. In 1836, when he was assigned to the provinces, he chose to give up his post with the government rather than leave the city that had become so important to him. (Part of the attraction was Mlle. Choquet, who became Leverrier's wife in 1837.) During the lean year that followed he taught at the second-rate Collège Stanislas, but in 1837 the position of *répétiteur* (assistant) to Gay-Lussac at the École Polytechnique became vacant and Leverrier's prospects brightened again.

Gay-Lussac's influence was decisive, and it was taken for granted that he would recommend one of his own students for the position. The most promising candidates were Leverrier and Henri Victor Regnault (1810–1878). Although Leverrier's papers had been well-received, Regnault's experiments on the action of sulfuric acid were more impressive. Regnault was also highly esteemed by Gay-Lussac's judicious colleague, Pierre Berthier (1782–1861). Still Gay-Lussac hesitated, reluctant to choose between two desirable men. At that critical moment another position, identical in rank, became available at the École Polytechnique, that of *répétiteur* in astronomy. Gay-Lussac knew that Leverrier spent much of his leisure time solving difficult problems in calculus and that he was fascinated with higher mathematics. Seizing the chance to secure two talented men for the faculty of the École Polytechnique, Gay-Lussac recommended Regnault for the post of *répétiteur* in chemistry and Leverrier for the same position in astronomy. The walls of journals separating one scientific specialty from another were not high in 1837, and Leverrier accepted the offer with aplomb. He gave no indication that the sudden change of fields in mid-career caused any internal conflict. The promising young chemist, "without regret as without effort, without dividing his attention and without looking back, detached himself from chemistry and, obedient to the decree of chance that pointed out his course, rapidly became an astronomer." [3] His new position had been held by the prominent geodesist Félix Savary (1797–1841), and Leverrier's primary concern was that he should be worthy of the post. A few days after accepting it he wrote to his father:

> In daring to accept a position held successively by Arago, Mathieu, and Savary, I incur the obligation of insuring that the post which they occupied does not diminish in public esteem, and in order to accomplish that I must not merely accept, but must actively seek out opportunities to improve my knowledge . . . I have already begun to mount the ladder [of success], why shouldn't I continue to climb? [4]

To Leverrier a prerequisite for the next rung seemed to be a thorough knowledge of celestial mechanics, and he purposefully set about acquiring it. It was his good fortune to be endowed with a mutually consistent set of psychological characteristics: his driving ambition was complemented by his unusual capacity for difficult and tedious work. Leverrier never balked at problems involving vast amounts of calculation or investigations that because of their sheer immensity might have deterred less industrious men. For his first essay in celestial mechanics the chemist-turned-astronomer chose to investigate the stability of the solar system.

As we have seen, this problem had been studied in the late eighteenth century by Laplace and Lagrange. Although Laplace had derived general stability equations for the solar system, he had calculated no limits for the periodic oscillations of the planetary orbits. The limits for any individual orbit could be found only by integrating huge differential equations of condition, equal in degree to the number of known planets. Lagrange broke these equations down by dividing the planets into two groups, each composed of bodies with relatively strong mutual attractions. The first group included Jupiter, Saturn, and Uranus; the second, Mercury, Venus, the Earth, and Mars. After deriving equations of condition for each of these systems, Lagrange described their mutual interactions in a third set of simpler formulas. His solution, though unique, was vitiated by the incorrect physical constants that he had to work with. (For example, he used a figure for the mass of Venus that was 50 percent too large.) By 1837 his results were no longer considered accurate.

In that year Leverrier began work on a completely new solution. He devised a more compact set of equations than those Lagrange had used and worked through the tedious computation of a new set of astronomical constants. On September 10, 1839, after two years of work, he submitted the paper containing his solution to the Académie des Sciences. This essay, which marked Leverrier's debut as a theoretical astronomer, was entitled "Sur les variations séculaires des orbites planetaires."

The first part of the paper dealt with the eccentricities and apsides of the planetary orbits. Leverrier found that the eccentricity of the Earth's orbit was slowly diminishing and would reach a minimum value of 0.0033 in 24×10^3 years. The eccentricity would then gradually increase, attaining a maximum value of 0.021 about 46×10^3 years later. (This was not the largest possible eccentricity of the terrestrial orbit, which Leverrier gave as 0.07775, but only the maximum reached in the specified cycle.) The orbit of the Earth's exterior neighbor, Mars, showed a very slow rate of change: it would require 1.8×10^6 years to reach its maximum eccentricity of 0.14224. About half that time was required for the combined Jupiter–Saturn–Uranus system to complete its cycle of oscillation, but the period of Saturn's orbital eccentricity was given as only 16.114×10^3 years. Leverrier was particularly proud of his calculation for the period of the three-planet system; he asserted that the error in his figure of 900,000 years did not exceed, 4,000 years, or 0.44 percent!

The second part of Leverrier's essay treated the secular variations in the inclinations and longitudes of the orbital nodes, and included limits for the inclinations of the planetary orbits. The final section consisted of tables that gave the elements of Mercury, Venus, the Earth, and Mars for a 200,000-year period centered at the epoch of 1800.

How much did Leverrier's results differ from those reached by Lagrange? In some cases, notably the formulas for the elements of the planets, they differed considerably. On the other hand, Leverrier found upper limits for the orbital eccentricities almost identical to those given by Lagrange, even though the two men had used very different values for the assumed masses of the planets. In explaining the similarities, Leverrier stated that the expressions for orbital eccentricities were nearly independent of variations in planetary mass. He also asserted that, if Lagrange had used any other (incorrect) value for the mass of Venus than the one he chose, his results would have been markedly different from Leverrier's. Lagrange had, according to Leverrier, used a figure that canceled its own error in the course of calculation. Leverrier's paper elicited much favorable comment: "[His] debut was brilliant . . . it heralded to the scientific world the appearance of a potential successor to Laplace." [5]

In 1840 Leverrier submitted to the Académie a second paper on the same general topic. This essay, which appeared in the *Connaissance des temps* for 1844 (published in 1841), contained two important innovations. The first was the discovery that equations for the secular variations of the planets could not be limited to first-power terms for eccentricities and inclinations. Because these functions changed so slowly, it had always been intuitively assumed that the omission of their higher powers would not alter the secular variations significantly. Leverrier tested this assumption by carrying his equations out to the third powers of the inclination and eccentricity. He found that, contrary to expectation, the higher powers were not negligible; their inclusion resulted in sensible alterations of the final values and important changes in the theories of Mercury, Venus, the Earth, and Mars. The second, and perhaps the most important, innovation in Leverrier's paper of 1840 was his revised form of equation. In their final state, his equations for the planetary motions allowed corrections to be directly inserted; for the first time an improved result could be reached by means of a simple substitution instead of hours of tedious computation.

Leverrier's analytical virtuosity attracted the attention of the dean of French astronomy, François Arago. Arago was director of the Paris Observatory and secretary of the Academy of Sciences; he also acted as a one-man steering committee for French astronomy. One of his most constructive practices was bringing important unsolved problems to the attention of talented young men. Arago inducted Leverrier into this scheme of intellectual logistics by suggesting that he work on the theory of Mercury.

Leverrier was a politic man; on this occasion, as on many later ones, he immediately dropped his own work and adopted the project suggested by those above him. Though he worked on the theory of Mercury for the next three years, he was unable to bring it to a satisfactory state. His memoir on the subject, "Détermination nouvelle de l'orbite de Mercure et de ses perturbations," was presented to the Academy in 1843. By that time, Leverrier had learned the value of intelligent advertising; in the introduction to the paper he smoothed the way for his analytical stalemate by pointing out the numerous difficulties encountered by previous investigators of the theory of Mercury. Leverrier had hoped to provide improved tables and methods for computing new ephemerides, but he could not accomplish his aims and left the investigation in tentative form. He did not complete it until sixteen years later.

Late in 1843 Leverrier became interested in comets of short period, and he worked on this subject until the summer of 1845. He spent most of the time investigating the comet of 1770, which had been the subject of some controversial hypotheses. This comet, though named after Lexell, was discovered by Charles Messier on June 15, 1770. On June 21 it was visible to the naked eye, and by July 1 it had approached to within 363 Earth radii. (For comparison, the distance of the Moon is about 60 Earth radii.) On July 4, it disappeared into the Sun's glare. Alexander Guy Pingré (1711–1796), a member of the Academy and director of the observatory at the Abbaye de Sainte Geneviève, calculated an ephemeris which predicted that the comet would soon be visible again. Messier rediscovered it on August 4; it was still bright enough to be seen without a telescope, and it remained so until August 26. Observers tracked the comet until the beginning of October, when it again disappeared.

Using a large number of observations, Pingré tried to compute a parabolic orbit for the comet, but he could not fit an equation to the data. The distinguished German physicist Johann Heinrich Lambert (1728–1777) attempted the same calculation, but he too was unable to find any suitable parabola. Only when Lexell tried to fit the observations without any preconceived notions did he find the true shape of the orbit. It was, surprisingly, an ellipse of very small dimensions, one in which the comet would have a period of only $5\frac{1}{2}$ years. The question was immediately raised, if the comet of 1770 had such a short period, why had this remarkably conspicuous object never been observed before? Lexell had no ready answer, but it was obvious that, if his hypothesis was correct, the comet would be seen late in 1776 and again in 1782. In both years astronomers waited in vain; the brilliant first appearance of the comet of 1770 was also its last.

Left in a difficult position, Lexell rechecked his calculations and found that to all appearances they were correct. In the course of the check he realized that according to his theoretical orbit the comet would have passed very close to Jupiter in 1767 and 1779. Seizing on this possibility, Lexell attempted to save his original hypothesis by proposing a more daring one. He suggested that the comet had traveled in an invisible orbit until it approached Jupiter in 1767. At that time the powerful attraction of Jupiter, acting at a short distance, had pulled the comet into a new orbit, the one in which it had been observed from the Earth. Lexell maintained that the comet had continued in this orbit until its next approach to Jupiter in 1779, when it had again been deflected into an orbit invisible to terrestrial observers.

This dramatic idea was not reexamined until 1806. In that year, Johann Burckhardt, acting on Laplace's suggestion, reviewed Lexell's calculations. He verified that they were correct, and, using formulas supplied by Laplace, he calculated the effect of Jupiter on a theoretical comet traveling in the orbit computed by Lexell. He concluded that the required deflection was perfectly feasible, and his results, published in the *Mécanique céleste,* were accepted with confidence by most astronomers.[6]

This was still the situation in 1844, when Leverrier began a new investigation of the comet of 1770. After reanalyzing the original data, he decided that the mean motion of the comet, on which the elements of its orbit depended, could not have been determined correctly from the observations made in 1770. This meant that the orbit originally assigned to the comet and Burckhardt's results of 1806 must have both been incorrect. Leverrier tried to determine the effect of the known error on the calculations for the comet's near approach to Jupiter in 1779. He found that there was a whole family of orbits in which Lexell's comet could have moved between 1770 and 1779. Following each of these paths analytically, Leverrier proved that the comet could not have been captured by Jupiter and remained a satellite. There were, however, other possibilities: the comet could have been deflected into a hyperbolic orbit, in which case it would never again be visible from the Earth, or it could have followed one of many different elliptical orbits. In the event the latter might have happened, Leverrier tabulated all the pertinent ellipses in order that the orbits of future comets could be readily compared with them and any possible identity confirmed.

On November 22, 1843, Hervé Faye (1814–1902), a young French astronomer, discovered a new comet. A short time later, when Leverrier's tables became available, it appeared that the orbit of Faye's comet might fit one of the tabulated ellipses for the comet of 1770. Benjamin Valz was very excited about this potential identity and advanced a complicated hypothesis to account for it. In Valz's scheme the attraction of Jupiter was made responsible for three markedly different, delicately administered perturbations. Jupiter appeared to have regained his throne in a Christian era, at least as *deus ex machina*.

Valz's hypothesis stimulated Leverrier to investigate the question himself. In order to decide about the identity of Lexell's and Faye's comets, Leverrier had to calculate the perturbations of the latter back to the year 1781. The errors in the elements of both comets made the comparison difficult and conclusions uncertain, but after much work Leverrier announced that, no matter what allowances were made, the two sets of elements could not be reconciled and the comets were therefore not identical.

The case of Faye's discovery had hardly been settled when, on August 22, 1844, Francesco de Vico (1805–1848), director of the observatory at the Collegio Romano, found another new comet. The hue and cry over cometary identities began again, and Leverrier went back to work. This time he had competition. Two Parisian astronomers, Paul Auguste Ernest Laugier (1812–1872) and Félix Victor Mauvais (1809–1854), believed that de Vico's comet was a reappearance of one observed by Tycho Brahe and Christoph Rothmann in 1585. Edmund Halley had calculated the orbit of Tycho's comet as a parabola, but Laugier and Mauvais decided that its path was actually an ellipse with a period of 5 years and 2 months. Although the places of perihelion of de Vico's and Tycho's comets differed by more than 30°, and their orbital nodes by 22°, their eccentricities were nearly the same. Laugier and Mauvais attributed the differences to perturbation effects, and asserted that the comets were identical.

Leverrier was partial to a different hypothesis: he thought that the comet of 1844 was the same as Lexell's, and set out to test his conjecture. Starting with elements calculated by Franz Friedrich Ernst Brünnow (1821–1891) for de Vico's comet, Leverrier traced its path back to 1781. He found that de Vico's comet, like Faye's, had passed very close to Jupiter, but in the year 1814. The resulting major perturbation could account for some, but not all, of the differences between its elements and those of the comet of 1770. Leverrier reluctantly took Lexell's comet out of the running, and went on to test the hypothesis advanced by Laugier and Mauvais. He found that, as he traced the theoretical orbit of de Vico's comet back in time, its nodes and perihelia receded farther and farther from those of Tycho's comet until, in 1585, they were respectively 180° and 338° apart. There was not the slightest possibility that the two comets were identical. In the course of his investigation Leverrier discovered that a comet observed by Philippe de La Hire (1640–1718) in 1678 showed remarkable similarities to the one seen by de Vico. After making a careful comparison he decided that La Hire and de Vico had probably observed different visits of the same comet.

When Leverrier finished this problem he was thirty-four years old; he had earned a reputation as a gifted analyst, and he seemed to have a special talent for tracking down elusive discrepancies in the elements of celestial bodies. He had already done some work on the perturbations of Uranus and during the summer of 1845, François Arago decided that he was the perfect man to investigate the problem more deeply. As before, Leverrier gave precedence to Arago's suggestion, and concentrated his attention on the puzzling motions of the seventh planet.

CHAPTER 4

. . . .

ADAMS

BY JUNE 1845, when Leverrier began to study the motion of Uranus, John Couch Adams, a young Fellow of Cambridge University, had already worked on the problem for two years. Adams was born on June 5, 1819, at Lidcot, a farmhouse in the parish of Laneast, Cornwall.[1] His father, Thomas Adams, was a tenant farmer like his forebears and a devout Wesleyan. He was also a devoted husband and father. John's mother Tabitha had a gentle disposition and some talent for music; she passed these attributes on to her eldest son. John Adams never forgot the secure childhood environment provided by his parents.

John first went to school in a farmhouse at Laneast. When he was eight years old he started attending a class taught by R. C. Sleep, who advertised himself as "Professor of Caligraphy, Stenography, French, Hebrew, etc. . . . Mr. Sleep Challenges any man in England for Caligraphy, Stenography or the Mathematics." Despite his boast, Sleep was not much of a mathematician. When he came to Laneast in 1827, he had only one ancient algebra text, which he had not mastered. After John Adams had studied with him for two years, the schoolmaster and his ten-year-old pupil set out to discover the mysteries of algebra together. John quickly outpaced his teacher.

In 1830 Thomas Adams took his precocious son to visit a relative named Pearse who lived at Hatherleigh in

Devon. During the visit John easily bested Pearse's son in a problem-solving contest. Pearse was so impressed that he arranged a competition between John and the local schoolmaster, who had a reputation as a mathematician. It is possible that the Hatherleigh teacher's mathematical talent, like Mr. Sleep's, existed mostly on his bill of advertisement; in any case, he was vanquished by John Adams, aged eleven. After witnessing John's second triumph, Pearse told Thomas Adams, "If he were my boy I would sell my hat off my head rather than not send him to college." It was also at Hatherleigh that John saw his first set of astronomical globes and was taught how to use them. The experience stimulated an interest in astronomy that was to last through his entire life.

It was obvious that John's extraordinary gift for mathematics should be nurtured. Thomas and Tabitha Adams had already concluded that their eldest son would never make a farmer. Encouraged by Pearse's comments, they decided to give John the best education they could afford, and hoped they might someday be able to send him to a university.

Early in 1831 John was removed from the academic care of Mr. Sleep and sent to Devonport, where Tabitha Adams's cousin, the Reverend John Couch Grylls, headed a better school. Grylls believed strongly in the value of classical education, but unfortunately, like his lesser contemporaries in Laneast and Hatherleigh, he did not attach much importance to the study of mathematics. As a result, John Adams's own intellectual curiosity was responsible for the mathematical supplements to his formal schooling. Grylls's school was near the Town Hall, which housed the classrooms and library of the Devonport Mechanics Institute, and it was here that John began his scientific education. He spent all of his spare time in the Institute library, poring over the astronomical articles in *Ree's Cyclopaedia* and working his way, unaided, through a copy of Vince's *Fluxions.* His preoccupation did not go unnoticed; in 1834 he was awarded a copy of J. F. W. Herschel's *Astronomy* as a school prize. He later regarded this incident as an encouragement of critical importance to his education.

John was fascinated with comets and eclipses. On October 17, 1835, he described to his parents his observation of Halley's comet: "You may conceive with what pleasure I viewed this, the first Comet I ever had a sight of, which at its visit 380 years ago threw all Europe into consternation, but which now affords the highest pleasure to astronomers for proving the accuracy of their calculations and predictions." The lunar eclipse of April 20, 1837, was the subject of John's modest first publication. On April 24, he wrote home:

> I observed the eclipse last Thursday with a small spyglass which I borrowed: the moon looked most delightful after the end of the eclipse. At the request of Mr. Bate, a young man of my acquaintance, who reports for the [Devonport] *Telegraph,* I wrote next morning a few lines on the eclipse, which were inserted in the paper the following day . . . Mr. Richards, the editor of the *Telegraph,* tells me that my article on the eclipse has been copied into several of the London papers.

Beginning about this time, the scientific notes in John's diary showed a marked increase in sophistication. The following entries, made in the spring of 1837, were typical:

> [April 1] Investigated for the first time the formulae of Olbers for finding apparent place of Moon in terms of true [heliocentric coordinates].

> [May 16] Read for first time the investigation of a sphere given in the Cyclopaedia Britannica and after I got into bed

solved the same problem by another method of my own, considering the globe as an assemblage of concentric spherical shells.

By this time it was clear that John should go to a university, and his studies were more and more directed to that end. Working especially hard on mathematics, and teaching himself, he finished the standard texts on conic sections, differential calculus, theory of numbers, theory of equations, and mechanics. If his own mathematical accomplishments were any indication, there were few teachers in England to equal him.

John's parents were much less worried about their son's academic qualifications than about their own ability to bear the heavy expenses of a college education. Two factors combined to help them meet the impending financial burden. The first was the death, in 1836, of Grace Couch, Tabitha Adams's adoptive mother, who left Tabitha some property and a small annual income. The second was that John began to earn money by tutoring the sons of local gentry during his school vacations.

In the spring of 1839, John began to study with the Reverend George Martin, a local curate who was a Cambridge graduate and a competent mathematician. It was probably because of his influence that Adams decided to apply to St. John's College of Cambridge University. Under Martin's direction, John worked intensively through the summer, preparing for the autumn sizarship examinations at Cambridge. Early in October he was ready to leave. His whole family saw him aboard the night coach which took him to London on the first stage of his journey.

John's mathematical talent and the excellence of his preparation were noted soon after he arrived at Cambridge. His generous and unaffected nature won him friends immediately. A. S. Campbell, who sat for the same entrance examinations, later described the mixed discouragement and admiration that he felt on first meeting Adams:

> He and I were the last two at a viva-voce examination; he broke the ice by asking me to come to his rooms to tea. I went and we naturally had a long talk on mathematics, of which I knew enough to appreciate the great talent of my new friend. I was in despair, for I had gone up to Cambridge with high hopes and now the first man I meet is something infinitely beyond me and whom it was hopeless to think of my beating. If there were many like him, my hopes of success were gone. A few days experiences soon relieved me and I knew what a wonderful man I had met. If I could keep *near* him in the examinations I should do very well.

As Campbell expected, Adams did brilliantly, winning both his sizarship (which carried a partial allowance toward college expenses) and the enthusiastic praise of the examiners.

Adams set very high standards for himself at Cambridge. His diaries show that he tried to schedule his work with great care, and also that he was sometimes unable to meet his own strenuous demands. On such occasions he reprimanded himself in his diary with entries such as, "Had a short paper but took a long time about it, having made myself very imperfectly acquainted with the subject," or "Had a paper in the Planetary Theory but was, as usual, unprepared." Certainly no one else would have accused him of being unprepared; he easily won the highest mathematical awards in his college, and became successively exhibitioner, prizeman, and scholar. He also took first prize in Greek Testament every year he remained at Cambridge.

During his first long vacation, Adams began tutoring other students. He enjoyed teaching, and the small additional income helped to relieve the burden on his parents. He spent most of his vacations in Cambridge, but returned to Lidcot for one month before the opening of each academic year. In June 1840 Adams and two friends walked to London (it took thirty hours) and spent a week there sightseeing. Not surprisingly, one of the places they went out of their way to visit was the Royal Observatory at Greenwich.

Astronomy was rapidly becoming John Adams's ruling passion. Early in 1841 he rebuked himself in his diary for favoring this preoccupation: "I have badly broken my plan today, chiefly wasting my time with Astronomy. I resolve not to let my astronomical amusements interfere with my regular work." In keeping this resolution he put off a much-anticipated first visit to the University Observatory until the Easter vacation of 1841. On this tour he was briefly shown the 12-inch Northumberland equatorial telescope, the mural circle, and the transit instrument. Later, during a second visit, Adams and Campbell were both allowed to inspect the instruments more closely. They had no premonition of the critical role the Northumberland telescope was to play in John Adams's future.

On June 26, 1841, Adams spent some time browsing in a Cambridge bookshop (Johnson's, on Trinity Street). While there he read George Biddell Airy's *Report* of 1831–1832 to the British Association for the Advancement of Science. This ten-year-old publication described Bouvard's dilemma of 1821 and his rejection of the ancient observations of Uranus in favor of the modern ones. It also mentioned that by 1831 Bouvard's tables were seriously in error.[2] For John Couch Adams this was a critical encounter; a week later, on July 3, 1841, he wrote the following memorandum on a slip of paper (still preserved in Adams's notebooks in the Library of St. John's College, Cambridge):

1841. July 3. Formed a design, in the beginning of this week, of investigating, as soon as possible after taking my degree, the irregularities in the motion of Uranus, wh. are yet unaccounted for; in order to find whether they may be attributed to the action of an undiscovered planet beyond it; and if possible thence to determine the elements of its orbit, &c approximately, wh wd. probably lead to its discovery.

Not long after, Adams was talking over his future plans with a fellow student named George Smith Drew. Drew said that after he had taken his degree he intended to go into the Church, and asked Adams what he was going to do. Adams answered, "You see, Uranus is a long way out of his course. I mean to find out why. I think I know." Drew was impressed by the decisiveness of Adams's reply; he later said it had given him a queer feeling, as though he had been favored with an oracle.

In 1841, Adams reached the point where he could indulge his passion for astronomy in his regular schoolwork. He took courses in lunar and planetary theory, and during the Michaelmas term he was first shown "the fixed lines in the spectrum." By the end of the year tension was already beginning to build up for a Cantabrigian ordeal, the mathematical tripos. This examination for baccalaureate honors in mathematical science was instituted at Cambridge early in the eighteenth century. By Adams's day it had become an ineradicable appraisal of scientific ability. The men who qualified for the highest class of tripos honors were known as wranglers, and were ranked numerically in order of performance, beginning with the senior (first) wrangler. Men in the second and third classes were called respectively senior and junior optimes. The examination usually consisted of eighteen three-hour papers and it encompassed both standard subjects ("bookwork") and special problems; the latter were original and extremely difficult. J. E. Littlewood, a well-known British mathematician, has pointed out that it was impossible to achieve a high rank by drill alone: "a man who did all the bookwork . . . and nothing else would have been about 23rd Wrangler out of 30."

After a year of practically uninterrupted preparation, Adams sat for the tripos of January 1843. Campbell, the apprehensive first acquaintance who became Adams's close friend, described part of the examination:

> In the Tripos examination I noticed that when everyone was writing hard Adams spent the first hour in looking over the questions, scarcely putting pen to paper the while. After that he wrote out rapidly the problems he had already solved in his head and ended by practically flooring the papers. Towards the end of the examination, while Adams and I were noting what we had done on the last problem-paper, Goodeve [afterwards professor of applied mechanics at the Royal School of Mines] looked over our shoulders. He was so staggered at Adams's record that he straightaway left Cambridge and did not put in an appearance for the papers of the last two days. In spite of this he came out ninth wrangler, but did not get his fellowship.[6]

Goodeve's appraisal was accurate. Not only was Adams senior wrangler—this was expected by everyone else if not himself—but he scored over 4000 marks in an examination in which the second wrangler made fewer than 2000. The result was that in 1843 there was a wider gap between the senior and second wranglers than between the latter and the "wooden spoon"—the student with the smallest number of marks. Campbell himself came out fourth wrangler and became a fellow of St. John's. Goodeve's despairing flight was indicative of the terrific emotional stress that the tripos examination could impose on sensitive or nervous individuals.

A few weeks after the tripos, Adams sat for and won the first Smith's prize, the highest mathematical trophy awarded by the University. This fresh triumph was followed almost immediately by his election as a fellow of St.

John's College. He was, as the letter announcing the good news to his parents showed, very happy:

> The list of Honours came out this morning and I hasten to remove the anxiety you have, no doubt, felt about the result of our examination. You may imagine what happiness I feel in being able to inform you that I have equalled the most sanguine expectations of my most sanguine friends, and have been favoured with the highest honours which this University can bestow. Very much of my gratification arises from the delight which I know my dear father and mother and all at home will experience in hearing it; and also from the reflection that I shall probably be now placed in a position to be of some service to my friends.

Adams considered that his best friends had been his parents, and his gratitude to them took precedence over everything else. Although he was eager to begin his long-deferred investigation of Uranus, he decided to allot only his vacations to the problem; he devoted all his spare time at Cambridge to teaching private pupils, and he sent the income from this work home to help repay his family and assist in the education of his younger brothers.

He had already told George and Thomas Jr. about his planned investigation and the hypothesis underlying it. Their references to it left little doubt about his convictions. On January 27, 1843, George wrote to John, congratulating him on his success in the tripos. Appended to the letter was a solicitous postscript: "I think you had better rest for a few weeks before you begin about the New Planet." Adams had obviously decided what was responsible for the irregularities of Uranus.

His interest in the problem had originally been provoked by a paper written in 1831, a time when, as we have seen, most astronomers had not yet accepted the solution of an external planet as the most probable one. In his memorandum of 1841 Adams had still been uncertain about the existence of a disturbing planet. Between that time and the beginning of 1843 he had somehow become convinced that this hypothesis was the only correct one. There is an interesting possibility that he may have acquired this conviction from one of Mrs. Somerville's books. Some years later, when he was asked about the origins of his investigation, Adams stated that his calculations were suggested by a sentence in the 1842 edition of *The Connexion of the Physical Sciences* which pointed out that the perturbations of Uranus might disclose the existence of an unseen planet.

Shortly after his election to fellowship, Adams outlined his proposed analysis to James Challis (1803–1882), the Plumian professor of astronomy at Cambridge. Challis encouraged him, lent him some books, and promised to supply whatever other necessary materials he could. In May 1843 Professor William Hallowes Miller also urged Adams to proceed with his plan. At the beginning of the next long vacation Adams returned to Lidcot and at last began the long-contemplated investigation of Uranus. George Adams described his brother's intense application to the problem:

> At home much of John's time during the vacation was occupied on long calculations. In the family he was always most cheerful and happy and he thoroughly enjoyed the country life; but he was never idle and did a large part of the work connected with his great problem at home. Night after night I have sat up with him in our little parlour at Lidcot, when all the rest had gone to bed, looking over his shoulder, seeing that he copied, added and subtracted his figures correctly, to save his doing it twice over . . . Often I have been tired and said to him "It's time to go to bed, John." His reply would be, "In a minute," and he would go on almost unconscious of anything but his calculations. In his walks at those times on Laneast Down, often with me, his mind would be fully occupied in his work. I might call his attention to some object and get a reply, but he would again relapse into his calculations.

THE DISCOVERY OF NEPTUNE

By October 1843 John Adams had arrived at a preliminary solution of the problem. Though approximate, it confirmed his belief that the irregularities in the motion of Uranus were caused by the perturbations of an exterior planet.

CHAPTER 5

. . . .

THE HYPOTHESIS

ADAMS was attempting to solve a problem of inverse perturbation: Given the disturbances of Uranus caused by an unknown planet, to find the latter's mass and elements, and hence its position at any time. The problem could be broken down into four parts: (1) determining the perturbation of Uranus for any given time; (2) resolving this perturbation into known and unaccounted-for components; (3) formulating equations in which the unidentified forces were related to the mass and elements of the hypothetical disturbing planet; (4) solving the equations for numerical answers.

Adams made two assumptions for his trial investigation. The first was the mean orbital distance of the supposed exterior planet. We have seen that in the beginning of the nineteenth century the discovery of the minor planets had established Bode's law in a position of scientific respectability. Adams used it without hesitation, assuming an orbital radius for his theoretical planet of 38.4 astronomical units, about twice the mean distance of Uranus from the Sun. His second assumption was that the unknown orbit probably had low eccentricity and could be approximated by a circle.

The first analysis was based only on "modern" observations. Adams used the tabular corrections given by Bouvard's equations of condition for the years 1781–1821; subsequent observations were taken from data published by Airy in the *Astronomische Nachrichten* and from the records of the Cambridge and Greenwich Observatories. The results of the exploratory calculations convinced Adams that his theory could be reconciled with observation.

A month after he finished his first solution Adams was distracted from his plans for a revised analysis by exciting news from France: on November 22, 1843, Hervé Faye had discovered a new comet. At Cambridge Professor Challis observed the comet on November 29, December 8, and December 16. From these observations Adams computed elements which Challis sent to the Royal Astronomical Society. On January 12, 1844, they were published in the Society's *Monthly Notices,* together with Adams's remarks about the comet. The latter were particularly interesting because Adams's conclusions were virtually identical to those reached independently by the French astronomers Valz and Leverrier. In a passage that would have sounded very familiar to the readers of the *Comptes rendus,* Adams suggested that

> the comet may, perhaps, not have been moving long in its present orbit, and that, as in the case of the comet of 1770, we are indebted to the action of *Jupiter* for its present apparition. In fact, supposing the . . . elements to be correct . . . the comet must have been very near *Jupiter* when in aphelion, and must have suffered very great perturbations, which may have materially changed the nature of its orbit.[1]

Faye's comet was not the last subject on which Adams and Leverrier were to publish similar independent opinions.

After this brief interruption, Adams resumed work on his second analysis of the problem of Uranus. In his first solution the greatest discrepancies of position had occurred between 1818 and 1826; these were precisely the years for which the fewest observations were available. Adams knew that the Greenwich planetary observations for this period were currently in the process of reduction, so he decided to take advantage of Professor Challis's offer of assistance and asked him to try to procure the unpublished data. This request marked the beginning of a closer relation between the two men.

Challis, like Adams, had been senior wrangler and first Smith's prizeman of his year at Cambridge (1825), and in 1826 he was elected a fellow of Trinity College. He was ordained in 1830, and in the following year he vacated his fellowship by marriage. On February 2, 1836, Challis was elected to succeed George Airy (who had been appointed Astronomer Royal) as Plumian Professor of Astronomy and Director of the Cambridge Observatory. A contemporary described him as "courteous in manner, kindly in disposition, simple and unassuming in character." His response to Adams's request for data was in keeping with this description.

On February 13, 1844, Challis wrote to Airy at the Greenwich Observatory:

> A young friend of mine, Mr. Adams, of St. John's College, is working at the theory of *Uranus,* and is desirous of obtaining errors of the tabular geocentric longitudes of this planet, when near opposition, in the years 1818–1826, with the factors for reducing them to errors of heliocentric longitude. Are your reductions of the planetary observations so far advanced that you could furnish these data? and is the request one which you have any objection to comply with? If Mr. Adams may be favoured in this respect, he is further desirous of knowing whether in the calculation of the tabular errors any alterations have been made in Bouvard's *Tables of Uranus* besides that of *Jupiter's* mass.[2]

Airy was very conscious of his service function as Astronomer Royal. He replied by return mail:

I send all the results of the observations of *Uranus* made with both instruments (that is, the heliocentric errors of *Uranus* in longitude and latitude from 1754 to 1830, for all those days on which there were observations, both of right ascension and of polar distance). No alteration is made in Bouvard's *Tables of Uranus,* except increasing the two equations which depend on *Jupiter* by 1/50 part.[3]

Adams received the supplementary data on February 16, but, though he was grateful to have them, he was at this time very busy. The pressure of his college duties soon mounted to the point where he felt obliged to set aside his investigation and several months elapsed before he was persuaded to work on it again by Professor William Miller. In the spring of 1844 Miller urged Adams to compete for the mathematical prize question on Uranus announced two years earlier by the Göttingen Academy of Sciences (page 56). Adams protested that he could not spare the time to prepare a comprehensive enough paper, but as a result of Miller's encouragement he finished blocking out his second analysis. His revised equations were to be based on a complete series of observations, including all of the known ancient positions:

For the modern observations, the errors of the Tables were taken exclusively from the Greenwich Observations as far as the year 1830, with the exception of an observation by Bessel, in 1823; and subsequently from the Cambridge and Greenwich Observations, and those given in various numbers of the *Astronomische Nachrichten.* The errors of the tables for the ancient Observations were taken from those given in the Equations of Condition of Bouvard's Tables.[4]

Adams again used a mean orbital radius of 38.4 astronomical units, but abandoned the simplification of circularity; he now took into account "the most important terms depending on the first power of the eccentricity of the disturbing planet."

In the autumn of 1844 Adams's work on Uranus was interrupted by an assignment from Challis, who like many other European astronomers had been observing de Vico's newly discovered comet. By the time Challis had collected enough data to determine an orbit for the comet, Adams had left Cambridge for his annual vacation at Lidcot. Challis sent his observations to Adams with a request that he compute the comet's elements. After several trial calculations, Adams concluded that the orbit was an ellipse of short period, but he was not too confident about this result and requested more data. On October 7, Challis sent back two later observations and suggested, "When you have satisfied yourself about the elliptic form of the orbit will it not be advisable to send a communication to *The Times* newspaper (the Astronomical Society is not sitting) in order that we may have the credit in England of first making the discovery." [5]

Adams followed this suggestion and sent his elements for de Vico's comet to the London *Times,* which printed them on October 15, but the publication did not accomplish the result Challis had hoped for. As we have seen, various French astronomers had become interested in de Vico's comet immediately after it was discovered, and it had been the subject of a masterly study by Urbain Leverrier. Adams did not know Leverrier, but he soon learned that Leverrier's conclusions about the comet were almost identical to his own. Although his letter to the *Times* was the first notice in England that the comet was periodic, the same conclusion had been announced earlier in France, as Adams informed his parents: "I found soon after my return [to Cambridge] that the French astronomers had obtained very nearly the same results as myself but earlier in time owing to their observations having been made earlier." This was the second time that Adams had independently attacked the same problem as had Leverrier, arrived at virtually the same result, and been beaten to publication. Except for his faintly chagrined note to his parents (and Challis's inferred disappointment), there were no apparent repercussions.

During the winter of 1844 and the following spring, Adams's duties as an assistant tutor of St. John's kept him too busy to pursue his private researches except during vacations. In June 1845 he attended the Cambridge meeting of the British Association for the Advancement of Science and became a life member. The meeting was an impressive experience for Adams, who had just turned twenty-six. On July 10 he described to his brother George "the pleasure of seeing for the first time some of our greatest scientific men, Herschel, Airy, Hamilton, Brewster, etc." The congregation of the great inspired Adams to new efforts. A few weeks after the meeting he wrote home that he had made further progress in his second attempt at "calculating the place of the supposed New Planet," and by the middle of September 1845 he had completed his new solution. He brought his figures for the mass and orbit of the hypothetical planet to Challis, who at once urged that they be sent to the Astronomer Royal at Greenwich. Since Adams was about to leave for his parents' home in Cornwall, he offered to deliver his results to Airy in person. Challis agreed to this, and on September 22 he wrote a letter of introduction to the Astronomer Royal:

> My friend Mr. Adams (who will probably deliver this note to you) has completed his calculations respecting the perturbation of the orbit of *Uranus* by a supposed ulterior planet, and has arrived at results which he would be glad to communicate to you personally, if you could spare him a few moments of your valuable time. His calculations are founded on the observations you were so good as to furnish him with some time ago; and from his character as a mathematician, and his practice in

calculation, I should consider the deductions from his premises to be made in a trustworthy manner. If he should not have the good fortune to see you at Greenwich, he hopes to be allowed to write to you on this subject.[6]

Before starting for Greenwich, Adams wrote out a summary of his results, which he left with Challis. Included in the summary was a significant addition to the elements of the hypothetical planet: its geocentric longitude on September 30, 1845. The function of this ordinarily omitted quantity was to locate an object for telescopic observation. Adams knew that, if the supposed planet existed, it would have a small but recognizable disk, one that the Northumberland telescope was easily capable of resolving. He obviously included the figure for geocentric longitude in the hope that Challis would attempt to search for the planet and identify it by its disk sometime during the last week in September.

Challis did not take the hint, and Adams ignored a more direct, and, as it later developed, equally crucial one himself. Before leaving Cambridge he showed his work on Uranus to Samuel Earnshaw, another fellow of St. John's, who suggested that he send a copy of his researches to the Cambridge Philosophical Society. Adams, probably setting his sights higher, ignored this advice and set out for Greenwich.

On his arrival at the Royal Observatory he was disappointed to find that Airy was in Paris attending a meeting of the French Institute. When the Astronomer Royal returned to Greenwich a week later, he found Challis's letter of introduction for Adams and replied to it immediately:

I was, I suppose, on my way from France, when Mr. Adams called here: at all events, I had not reached home, and therefore, to my regret, I have not seen him. Would you mention to Mr. Adams that I am very much interested with the subject of his investigations, and that I should be delighted to hear of them by letter from him? [7]

On October 21, 1845, Adams again tried to deliver the results of his investigation to Airy in person. On his way back to Cambridge from Cornwall he stopped off in London, where, "After waiting till about two o'clock for Thomas, I left for Greenwich to call on Mr. Airy who was unfortunately not at home. I left a note for him, however, containing a short statement of the results at which I had arrived."

Adams actually made two calls on Airy that day. The first time he left his card with his results and said that he would call again later; the card was taken to Mrs. Airy, but by some mischance she was not told of the caller's intention to return. When Adams did come back, Airy's butler answered the door and told him that the Astronomer Royal was at dinner and could not be disturbed. (It should perhaps be mentioned that Airy, because of his doctor's insistence, dined at precisely 3:30 every afternoon, an hour considered unusual even in his own day.) Airy's overprotective butler never informed him of Adams's second visit, and Adams returned to Cambridge feeling distinctly rebuffed. His hurt feelings and the circumstances that prevented him from meeting Airy were to have unhappy consequences for the prestige of British astronomy.

The paper that Adams left at the Royal Observatory was a concise summary of his solution to the problem of Uranus:

According to my calculations, the observed irregularities in the motion of *Uranus* may be accounted for by supposing the existence of an exterior planet, the mass and orbit of which are as follows:—

Mean Distance (assumed nearly in accordance with Bode's law)	38.4
Mean Sidereal Motion in 365.25 days.	1°30′.9

Mean Longitude, 1st October, 1845..........	323°34′
Longitude of Perihelion....................	315°55′
Eccentricity	0.1610
Mass (that of the Sun being unity)	0.0001656

For the modern observations I have used the method of normal places, taking the mean of the tabular errors, as given by observations near three consecutive oppositions, to correspond with the mean of the times; and the Greenwich observations have been used down to 1830: since which, the Cambridge and Greenwich observations, and those given in the *Astronomische Nachrichten* have been made use of. The following are the remaining errors of mean longitude:—

Observation — Theory

1780	+0″.27	1801	−0″.04	1822	+0″.30
1783	−0.23	1804	−0.04	1825	+1.92
1786	−0.96	1807	−0.21	1828	+2.25
1789	+1.82	1810	+0.56	1831	−1.06
1792	−0.91	1813	−0.94	1834	−1.44
1795	+0.09	1816	−0.31	1837	−1.62
1798	−0.99	1819	−2.00	1840	+1.73

The error for 1780 is concluded from that for 1781 given by observation, compared with those of four or five following years, and also with Lemonnier's observations in 1769 and 1771.

For the ancient observations, the following are the remaining errors:—

Observation — Theory

1690	+44″.4	1750	−1″.6	1763	−5″.1
1712	+6.7	1753	+5.7	1769	+0.6
1715	−6.8	1755	−4.0	1771	+11.8

The errors are small, except for Flamsteed's observation of 1690. This being an isolated observation, very distant from the rest, I thought it best not to use it in forming the equations of condition. It is not improbable, however, that this error might be destroyed by a small change in the assumed mean motion of the planet.[8]

As Adams hoped, a small change in the mean motion of his hypothetical planet reduced the error of 1690 to about 9″.

Although he tried to conceal it, Airy had a strong negative reaction to this paper. His attitude later turned out to be of critical importance, and in order to understand it we must know something about his background.

George Biddell Airy was born at Alnwick in Northumberland on July 27, 1801. He was a small, markedly asocial child who spent most of his time on schoolwork. When he was twelve years old he learned double-entry bookkeeping and, as we shall see, he attached a special and distorted value to this system throughout his life. In October 1819, Airy entered Trinity College, Cambridge, as a sizar. While there he worked extremely hard and earned the reputation of an unusually sober and conceited young man. He partially justified his high opinion of himself in 1823, when he came out of the mathematical tripos as senior wrangler. Shortly after he won the first Smith's prize and was elected a fellow of Trinity.

Airy's persistent industry won him first the Plumian professorship, in 1828, and then, in 1835, the post of Astronomer Royal. Throughout his career at Greenwich, his policies were dominated by two ideas: practicality and order.

> His nature was eminently practical, and any subject which had a distinctly practical object, and could be advanced by mathematical investigation, possessed interest for him. And his dislike of mere theoretical problems and investigations was proportionately great. He was continually at war with some of the resident Cambridge mathematicians on this subject.[9]

Airy's insistence on mathematical "practicality" was mild, compared to his obsession with order. The latter could only be regarded as pathological.

> The ruling feature of his character was undoubtedly Order. From the time that he went up to Cambridge to the end of his

life his system of order was strictly maintained. His accounts were perfectly kept by double entry throughout his life, and he valued extremely the order of book-keeping . . . He seems not to have destroyed a document of any kind whatever: counterfoils of old cheque-books, notes for tradesmen, circulars, bills, and correspondence of all sorts were carefully preserved in the most complete order from the time that he went to Cambridge; and a huge mass they formed. To a high appreciation of order he attributed . . . his command of mathematics, and sometimes spoke of mathematics as nothing more than a system of order carried to a considerable extent . . . he had the greatest dread of disorder creeping into the routine work of the [Greenwich] Observatory, even in the smallest matters. As an example, he spent a whole afternoon in writing the word "Empty" on large cards, to be nailed upon a great number of empty packing boxes, because he noticed a little confusion arising from their getting mixed with other boxes containing different articles; and an assistant could not be spared for this work without withdrawing him from his appointed duties.[10]

Airy was an extreme perfectionist, and he divided the people around him into two groups: those who had succeeded and were worthy of cultivation, and those who had not succeeded and were beneath consideration. Young people fell almost automatically into the latter group. Airy affected to have no confidence in their abilities, but in fact he determinedly blocked every opportunity for his own young assistants to demonstrate their talents. A contemporary wrote that Airy's despotism "militated—was almost avowedly intended to militate—against the growth of real zeal and intelligence in the staff [of Greenwich Observatory], and necessarily occasioned labour and discomfort out of proportion to the results obtained." [11] His refusal to delegate responsibility to younger men was sometimes carried to ridiculous extremes:

As an example of the . . . detail of the oversight which he exercised over his assistants, it may be mentioned that he drew up for each one of those who took part in the Harton Colliery experiment, instructions, telling them by what trains to travel, where to change, and so forth, with the same minuteness that one might for a child who was taking his first journey alone; and he himself packed up soap and towels with the instruments lest his astronomers should find themselves, in [County] Durham, out of reach of these necessaries of civilization.[12]

To George Biddell Airy—a man strongly opposed to theoretical investigations and skeptical of the abilities of younger scientists—John Couch Adams, aged twenty-six, delivered a paper that claimed to prove the existence of an undiscovered planet by a theoretical analysis of its effects on Uranus. Not surprisingly, Adams's hypothesis had a very cool reception at the Royal Observatory.

The degree to which Airy resisted Adams's claims is illustrated by an incident which did not become public knowledge until many years later. In October 1845, shortly after receiving Adams's results, Airy casually showed them to the Reverend William Rutter Dawes (1799–1868), a prominent amateur astronomer. In contrast to Airy's apathy, Dawes was tremendously impressed by Adams's work and wrote "at once" to his friend William Lassell (1799–1880), who was at that time installing the largest telescope in England (a 24-inch-aperture reflector) in his private observatory near Liverpool. Dawes sent Lassell Adams's positional data and urged him to look for a star with a disk in the region of the sky containing Adams's predicted place. When the letter arrived, Lassell was confined to bed with a sprained ankle, but he resolved to search when he recovered and placed the note on his writing table. Sometime later, when he looked for Adams's data, he found that a maid had removed and destroyed the letter, and, owing to the pressure of other work, he never undertook the search.

Airy had his own opinions about the feasibility of such an investigation, and he had formed them long before 1845. They were perfectly in keeping with his general mistrust of theoretical astronomy. In 1834, when the Reverend T. J. Hussey had written to Airy about the possibility of detecting a trans-Uranian planet (page 49), Airy had replied,

> I have often thought of the irregularity of *Uranus,* and since the receipt of your letter have looked more carefully to it. It is a puzzling subject, but I give it as my opinion, without hesitation, that it is not yet in such a state as to give the smallest hope of making out the nature of any external action on the planet . . . But [even] if it were certain that there were any extraneous action, I doubt much the possibility of determining the place of a planet which produced it. I am sure it could not be done till the nature of the irregularity was well determined from several successive revolutions.[13]

That is, Airy meant, in several hundred years. Eugène Bouvard was dissuaded from the same hope in 1837. In reply to Bouvard's statement of his intent to analyze the movements of Uranus for the effects of an exterior planet, Airy wrote:

> I think that, probably, you would gain much . . . by waiting . . . before you proceed with that part of your labour . . . With respect to the errors of the tables of *Uranus,* I think you will find that . . . [they] are increasing with fearful rapidity . . . I cannot conjecture what is the cause of these errors, but I am inclined, in the first instance, to ascribe them to some error in the perturbations . . . If it be the effect of any unseen body, it will be nearly impossible to ever find out its place.[14]

Adams's solution of the problem of inverse perturbation was thus a direct contradiction of Airy's considered opinion. The Astronomer Royal's negative feelings were indicated by the unusually long time he waited before replying. Airy habitually answered his correspondence by return mail; in this instance he delayed two weeks before writing to Adams on November 5, 1845. Perhaps it took that long for him to find some suitable scientific basis for his rejection. He found one, characteristically, in his own past work. In 1838 he had reported to the *Astronomische Nachrichten* that the tabular radius vector of Uranus was too small. In 1845 he expanded this fact into a scientific roadblock: "I therefore considered that the trial, whether the error of radius vector would be explained by the same theory which explained the error of longitude, would be truly an *experimentum crucis.*" [15] He wrote to Adams:

> I am very much obliged by the paper of results which you left here a few days since, shewing the perturbations on the place of *Uranus* produced by a planet with certain assumed elements. The latter numbers are all extremely satisfactory: I am not enough acquainted with Flamsteed's observations about 1690 to say whether they bear such an error, but I think it extremely probable.
>
> But I should be very glad to know whether this assumed perturbation will explain the error of the radius vector of *Uranus.* This error is now very considerable.[16]

Airy's repeated use of the word "assumed" in this letter indicated that he either did not realize, or was loath to admit, that Adams had not assumed his elements, but derived them from the perturbations of Uranus. In any case, Adams was nonplussed by the Astronomer Royal's reply. He never answered it, primarily because he felt that Airy was only putting him off by interposing the objection of the radius vector. Some time later, when J. W. L. Glaisher asked Adams why he had not replied to Airy's question, Adams said, "I should have done so; but the enquiry seemed to me to be trivial."

Since the whole matter was later regarded as anything but trivial, it will be followed out in some detail here. On

November 18, 1846, Adams wrote to Airy explaining his earlier position with regard to the radius vector:

> I need scarcely say how deeply I regret the neglect of which I was guilty in delaying to reply to the question respecting the radius vector of Uranus in your note of November 5, 1845. In palliation, though not in excuse of my neglect, I was not aware of the importance which you attached to my answer on this point, and I had not the smallest notion that you felt any difficulty on it . . . For several years past the observed place of Uranus has been falling more and more behind its tabular place. In other words the real angular motion of Uranus is considerably slower than that given by the tables. This appeared to me to show clearly that the tabular radius vector would be considerably increased by any theory which represents the motion in longitude, for the variation in the second member of the equation
>
> $$r^2 \cdot d\theta/dt = \sqrt{a(I - e^2)}$$
>
> is very small. Accordingly I found that if I simply corrected the elliptic elements, so as to satisfy the modern observations as nearly as possible, without taking into account any additional perturbations, the corresponding increase in the radius vector would not be very different from that given by my actual theory . . . I was . . . much pained at not having been able to see you when I called at the Royal Observatory the second time, as I felt that the whole matter might be better explained by half an hour's conversation than by several letters.[17]

Adams's letter did not conceal his underlying feelings; a year after the event, he still smarted from his failure to see Airy at Greenwich.

Challis brusquely dismissed Airy's objection about the radius vector, on December 17, 1846: "It is quite impossible that [Uranus's] longitude could be corrected during a period of at least 130 years independently of the correction of the radius vector . . . the investigation of one correction necessarily involves that of the other." [18] Two days later Airy wrote to Challis defending his position:

There were two things to be explained, which might have existed each independently of the other, and of which one could be ascertained independently of the other: viz. error of longitude and error of radius vector. And there is no *a priori* reason for thinking that a hypothesis which will explain the error of longitude will also explain the error of radius vector. If, after Adams had satisfactorily explained the error of longitude he had (with the numerical values of the elements of the two planets so found) converted his formula for perturbation of radius vector into numbers, and if these numbers had been discordant with the *observed* numbers of discordances of radius vector, then *the theory would have been false, not* from any error of Adams's *but* from a failure in the law of gravitation. On this question therefore turned the continuance or fall of the law of gravitation.[19]

This was not a widely shared opinion. Airy and Adams were arguing from different vantage points, and with different basic assumptions. If Airy's demand for a circular proof was a product of his psychological bias and his obsession with numerical results, Adams's assumption was mathematically incorrect. As J. E. Littlewood has pointed out, in the equation that Adams sent to Airy the variation in the angular momentum ($r^2 \cdot d\theta/dt$) was proportionately of the same order as the error in angular position ($\theta =$ longitude); correction of the longitude did not automatically correct the radius vector.[20]

Whatever the validity of his objections, Airy had managed to discourage Adams. The Astronomer Royal did not, of course, make any effort to find out why Adams had not replied to his letter. With more self-righteousness than regret, he judged that "Adams's silence . . . was so far unfortunate that it interposed an effectual barrier to all further communication. It was clearly impossible for me to write to him again." [21]

Airy's letter to Adams was dated November 5, 1845. On November 10, Urbain Leverrier presented his "Premier mémoire sur la théorie d'Uranus" to the French Academy of Sciences.[22] In this paper Leverrier announced the results of a completely new analysis of the motion of Uranus. He based his work on observations made at the Greenwich Observatory from 1781 to 1801 and at the Observatoire Royal in Paris from 1801 to 1845. Leverrier used two different systems of analysis, the standard one from Laplace's *Mécanique céleste* and a method of his own devising that yielded simultaneous perturbations. His whole treatment was rigorous to an unprecedented degree: "I have not omitted any inequality of more than a *twentieth* of a second of arc. No perturbation has been neglected because it was presumed to be insensible. Everything has been determined with the same rigor." Leverrier's results could be applied to any epoch. At the end of his abstract he set up hypothetical elements for Uranus based on the observations made between 1790 and 1820 and showed that, taking all known factors into account, tables calculated from these elements would show an error of more than 40″ of arc at the opposition of 1845. "Such," he concluded, "is the order of the error in the actual tables, except that the discrepancy is larger, and the surplus can be attributed to outside causes, whose effect I will evaluate in a second Memoir."

Leverrier's paper reached England in December 1845. Airy regarded it as "a new and most important investigation." [23] In the Astronomer Royal's opinion, only after the publication of Leverrier's memoir could it "be truly said that the theory of *Uranus* was now, for the first time, placed on a satisfactory foundation."

This was the situation at the end of 1845. In England, John Couch Adams had solved his problem and submitted the result to George Airy, who had pigeonholed it at Greenwich and (intentionally or not) rebuffed its author. On November 14 Adams had been elected a fellow of the

Royal Astronomical Society. In France, Urbain Leverrier had published his first memoir on the theory of Uranus, which had made a great impression on his Parisian colleagues and on the British Astronomer Royal. On January 19, 1846, Leverrier was elected to membership in the astronomy section of the French Institute.

CHAPTER 6

. . . .

THE DISCOVERY

ONE of the least-expected consequences of Leverrier's first memoir on Uranus was the complete discrediting of Alexis Bouvard. Leverrier's painstaking review of the *Tables of Uranus* showed them to be a shoddy patchwork of errors. Bouvard's equations of condition were analytically and numerically incorrect, the three methods he gave to determine the orbital eccentricity of Uranus yielded three widely different answers, and his first two tables were in complete disagreement on the secular motion of the mean longitude. The finished work contained an extraordinary number of typographical errors. (George Airy independently discovered and published many of these errors at about the same time.) By the time Leverrier had finished correcting mistakes and taking previously ignored factors into account, he had some hope of their combined effects reconciling the discrepancies of the tables. To test this possibility, he set up new equations of condition based on the corrected tables and used them to predict the position of Uranus, allowing observational errors of $\pm 4''$ in the modern positions and $\pm 15''$ in Flamsteed's observations. Even by combining the maximum values of all errors to reduce the discrepancy, Leverrier could subtract only $92''$ from a total difference of $356''$, leaving more than $4'$ of arc unaccounted for. It was this test which convinced him that some unknown perturbative force was acting on Uranus.

Leverrier spent the winter and spring of 1846 working on his promised second memoir. On June 1 he presented his "Recherches sur les mouvements d'Uranus" to the Academy of Sciences.[1] In this paper he proposed "to examine the nature of the irregularities in the motion of Uranus; to trace them to their source, while seeking to discover . . . the direction and magnitude of the force which produces them."

Leverrier attacked the problem by a process of elimination. Having first satisfied himself that no known planet was causing the irregularities, he examined the other possible explanations for the strange behavior of Uranus: a resisting ether; an undiscovered massive satellite of Uranus; an altered law of gravitation; a crucially timed collision with a comet; an unknown planet. After careful analysis he rejected the first four of these hypotheses. There was one article of faith among the discards; Leverrier asserted that an altered law of gravitation was for him "a last resort to which [he] would not turn until all other potential causes for the discrepancies had been investigated and rejected." Deferring to his Newtonian allegiance, he concluded that the irregularities in the motion of Uranus were perturbations caused by an undiscovered planet. (It should be noted that on any unmagnified scale these perturbations were very small. If, in 1844, the real and theoretical (tabular) Uranus, 2′ apart, could have been placed side by side in the sky, they would have appeared to even the sharpest naked eye as a single planet.)

The next step was to locate the disturbing body. Leverrier showed that the unknown planet could not be within the orbit of Saturn. If it were, its perturbation effects would be greater on Saturn than on Uranus, and there were no unaccounted-for perturbations of Saturn. "Suppose," he continued, "the planet were situated between the orbits of Saturn and Uranus?" It would then, by the same reasoning, have to be closer to Uranus than to Saturn, and of small enough mass to cause only perturbations of the observed magnitude. Leverrier showed that, since its period was determined within narrow limits by Kepler's third law, such a small hypothetical planet could not be responsible for the slow disturbances of Uranus. "Therefore," he reasoned, "the perturbing planet will be situated beyond Uranus."

How far beyond Uranus? Leverrier was not above using Bode's law, although he avoided mentioning it by name. A Gallic reflexiveness marked his references to "the remarkable law which has revealed itself in the mean distances of the planets from the Sun." Following this rule, he assumed that the hypothetical planet had a mean orbital radius of 38 astronomical units, about twice the distance of Uranus from the Sun. He supported his assumption with a logical demonstration that: (1) the disturbing body could not be very close to Uranus, even though beyond it, for the reasons given above; (2) on the other hand the planet could not be as distant as three times the orbital radius of Uranus, for in order to cause the observed perturbations at such a distance its mass would have to be so large that its effect on Saturn would be comparable to that on Uranus. Leverrier also pointed out that, since the orbits of Jupiter, Saturn, and Uranus were inclined only slightly to the ecliptic, the same condition could be assumed in a first approximation to the orbit of the hypothetical planet.

Leverrier needed one more coordinate, the longitude at any epoch, to determine the position of the disturbing planet on its orbital path. He found it by manipulating his equations of condition to yield a formula in which the mean longitude of the hypothetical planet was expressed in terms of its mass. He then calculated values of mass for each of 40 longitudes at 9-degree intervals from 0° to 360°. Any negative mass was, of course, inadmissible, as was any mass large enough to affect the movements of Saturn. Their successive elimination narrowed the possibilities until "there was only one region of the ecliptic where the perturbing planet could be located in order to account for the irregularities of Uranus; the mean longitude of the planet on the 1st of January, 1800, must have been between 243° and 252°." [2]

Leverrier defined this result more sharply by minimizing the residual differences from observation in his formula. After correcting for epoch, he asserted that the heliocentric longitude of the disturbing planet on January 1, 1847, would be 325°. He was sure that the error was less than 10°. Leverrier concluded the memoir with a reminder of the formidable difficulties he had overcome and a reference to Clairaut's speculation of 1758 about unknown planets (page 49): "Let us only hope that the stars of which Clairaut spoke will not all be invisible, so that, as in the chance discovery of Uranus, we will succeed in sighting the planet whose position I have just given."

The collective reaction of French astronomers to this memoir was little different from their reception of Leverrier's first essay on Uranus. There were some dignified huzzahs for another analytical "triumph," but no French observer made the slightest move to look for the hypothetical planet. The person most pleased with Leverrier's work, to judge by his own statements, was George Airy. He received Leverrier's second memoir on Uranus on June 23 or 24, 1846, and after examining it he wrote: "I cannot sufficiently express the feeling of delight and satisfaction which I received from it." [3] His delight might have been attributed to the fact that Adams's and Leverrier's results agreed within nearly 1° of Arc. Actually, it was an expression of Airy's insistent partiality to Leverrier and his corresponding skepticism of Adams:

> To this time I had considered that there was still room for doubt of the accuracy of Mr. Adams's investigations; for I think that the results of algebraical and numerical computations, so long and so complicated as those of an inverse problem of perturbations, are liable to many risks of error in the details of the process.[4]

Airy never neglected his doubts. On June 26, 1846, he wrote to Leverrier and asked whether the fact that the tabular radius vector of Uranus was too small could be "a consequence of the disturbance produced by an exterior planet, now in the position which you have indicated? I imagine that it would not be so." Airy's letter did not mention that John Couch Adams had left results substantially identical to Leverrier's at the Royal Observatory eight months earlier. Three days after he wrote to Leverrier, Airy attended a special meeting of the Board of Visitors to Greenwich Observatory. Both Challis and John Herschel were present when the Astronomer Royal spoke in support of a proposal to assign specific areas of the sky to individual observatories for study. He cited as an inducement "the extreme probability of now discovering a new planet in a very short time, provided the powers of one observatory could be directed to search for it."[5] He was convinced of this probability, he said, because of "the very close coincidence between the results of Mr. Adams's and M. Le Verrier's investigations of the place of the supposed planet disturbing Uranus."

Herschel was most impressed; as we shall see, he later publicized Airy's remark in a much more dramatic form. Challis, on the other hand, suffered a complete reversal of attitude. What confidence he had in Adams's theoretical work now seemed to be sapped, and he entirely lost the resolve shown earlier when he had urged Adams to publish his comet researches. His reaction was probably due to a personality conflict with Airy. The Astronomer Royal had missed few opportunities to criticize Challis's work publicly. Even Challis's letters on scientific subjects to the *Athenaeum* were frequently followed up by Airy's critical notes. Despite the possibility that this was in the best objective spirit, it is difficult to believe that Challis was not disturbed by Airy's continuous carping. Challis had championed Adams's cause at a time when Airy had been unreceptive to it. Now, after letting Adams's results gather dust for eight months, Airy began to represent himself as the source of encouragement in the search for a new planet. Challis was almost certainly put off by Airy's opportunism; after this time he followed the Astronomer Royal's assumed leadership with a passivity that mystified his contemporaries.

On July 1, Airy received Leverrier's reply to his query about the radius vector. Leverrier did not mince words; he wrote that in the course of correcting the true longitude "the radius vector is [automatically] corrected, without having to be dealt with independently. Excuse me, Sir, for insisting on this point." [6] Having answered the Astronomer Royal's question, Leverrier asked for some action in return: "If I may be allowed to hope that you will have enough confidence in my work to search for the planet in the sky, I will hasten, Sir, to send you its exact position as soon as I have [calculated] it." Airy was tremendously impressed by the self-assurance of this reply; it almost seemed as if his objections were satisfied more by Leverrier's positive tone than by his mathematical reasoning. The Astronomer Royal had "no longer any doubt upon the reality and general exactness of the planet's place." Despite this newfound conviction, he was no more eager to help find the planet than the French astronomers had been. He turned aside Leverrier's offer of precise coordinates with the excuse that "[his] approaching departure for the continent made it useless to trouble M. Le Verrier with a request for the more accurate numbers." Leverrier's offer was written on June 28; Airy was scheduled to leave for the Continent on August 10.

Since June 10, Airy had been host to Peter Andreas Hansen, director of the Seeberg Observatory. Hansen was deeply interested in perturbation theory, and, as we have seen (page 50), had corresponded with Alexis Bouvard in the early 1830's about the problem of Uranus. It is very probable that Airy and Hansen discussed Leverrier's paper, but an encounter that sharply delineated Airy's attitude toward Adams indicated that Hansen was not told about the coincidence of Adams's and Leverrier's results. On July 2, 1846, Airy and Hansen visited Cambridge University. Before this time, Airy had met John Adams only once, with Challis; the meeting must have taken place between December 4 and 6, 1845, but Airy "totally forgot where." Now, on St. John's bridge, Airy and Hansen met Adams by chance. One might have expected the Astronomer Royal to take advantage of this golden opportunity to publicize Adams's work, perhaps to arrange a meeting. Airy did neither of these things. He was preoccupied and cool; the only thing he later remembered about the interview was that it "might [have lasted] two minutes." His otherwise excellent memory invariably failed him in matters involving Adams.

It is possible that Airy's close brush with his conscience may have been partly responsible for his actions a week later. On July 6, he went to visit George Peacock, the

Lowndean professor of geometry and astronomy at Cambridge; Peacock's questions about the problem of Uranus, following close on Airy's chance encounter with Adams, overcame the Astronomer Royal's last resistance to the step he had been avoiding so determinedly. On July 9, 1846, Airy wrote to Challis, asking him to begin a search for the hypothetical planet. The Astronomer Royal's letter began with a phrase that was flatly unbelievable to anyone familiar with his past actions:

> You know that I attach importance to the examination of that part of the heavens in which there is . . . reason for suspecting the existence of a planet exterior to *Uranus.* I have thought about the way of making such examination, but I am convinced that (for various reasons, of declination, latitude of place, feebleness of light, and regularity of superintendence) there is no prospect whatever of its being made with any chance of success, except with the Northumberland Telescope.
>
> Now I should be glad to ask you, in the first place, whether you could make such an examination?
>
> Presuming that your answer would be in the negative, I would ask, secondly, whether, supposing that an assistant were supplied to you for this purpose, you would superintend the examination?
>
> You will readily perceive that all this is in a most unformed state at present, and that I am asking these questions almost at a venture, in the hope of rescuing the matter from a state which is, without the assistance that you and your instruments can give, almost desperate. Therefore I should be glad to have your answer, not only responding simply to my questions, but also entering into any other considerations which you think likely to bear on the matter.
>
> The time for the said examination is approaching near.[7]

Airy was beginning to worry. His next letter to Challis, written on July 13, had a distinctly urgent tone:

> I have drawn up the enclosed paper, in order to give you a notion of the extent of work incidental to a sweep for the possible planet.

I only add at present that, in my opinion, the importance of this inquiry exceeds that of any current work, which is of such a nature as not to be totally lost by delay.[8]

When Airy's letters reached Cambridge Challis was away on a trip. On July 18, soon after he returned, he replied to the Astronomer Royal that he had "determined on sweeping for this hypothetical planet . . . With respect to your proposal of supplying an assistant I need not say any thing, as I understand it to be made on the supposition that I decline undertaking the search myself . . . I purpose to carry the sweep to the extent you recommend." Airy later claimed that Challis had misunderstood his offer of an assistant. The offer was made in the belief that the Cambridge Observatory was already overloaded with work, and that an examination of the sort Airy proposed "would be entirely beyond the powers of its personal establishment."

If the Astronomer Royal believed this, and his resistance to a search had finally broken down, why did he not undertake it himself? There were two reasons. The first was his irrational dread of any alteration in the routine of the Greenwich Observatory. The second, announced publicly, was his belief that, if the planet did exist, it would be visible only in large telescopes. The largest telescope at the Royal Observatory in 1846 was the Sheepshanks equatorial refractor, which had an aperture of only 6.7 inches. It was vastly exceeded in light-gathering power by the Cambridge Observatory's 11.75-inch Northumberland telescope. (The object glass of this telescope, which was made by R. A. Cauchoix, was bought for the Cambridge Observatory by Hugh Percy, third Duke of Northumberland. The Duke also paid for mounting and housing the telescope, which was completed in 1839 under the supervision of George Airy.)

The plan of observation outlined in Airy's letter of July 13 was better suited to the construction of a star map than to the identification of a new planet. Although the Astronomer Royal claimed to be relieved of all doubts about the planet's location, he did not advise Challis either to look first where Adams and Leverrier pointed or to identify the hypothetical planet by its disk. Instead, Challis was to "sweep over, three times at least, a zodiacal belt 30° long and 10° broad, having the theoretical place of the planet at its centre; to complete one sweep before commencing the next; and to map the positions of the stars." [9] Airy calculated that the three sweeps would take 300 hours of observing. He further recommended that the telescope be held fixed and the stars allowed to drift across it, exactly as in a transit observation. (The Astronomer Royal had an almost religious personal preference for this method of observing.)

Challis was advised to use a magnification of at least 120. He found 166 power the most comfortable; at this magnification, the Northumberland telescope had a field of view of 9′ of arc. Airy had estimated that with a field of 15′ of arc at least 240 individual passes would be necessary. Furthermore, although Adams had assured Challis that the planet would appear no smaller than a star of the ninth magnitude, Challis elected to map all stars down to the eleventh magnitude; he thought that in this way the search would be "less liable to ultimate disappointment." There were more than 3000 stars brighter than the eleventh magnitude in the zone Airy had defined. In retrospect, it seems almost incredible that Airy's plan was adopted. It was like searching for a particular bright pebble on a beach by removing, one by one, thousands of other pebbles from a large area around the point where (one had been told) the desired object was lying.

Late in July 1846, when Challis told him of the projected search, Adams calculated the right ascension and declination of the planet for every twentieth day between July 20 and October 8, and for every fifth degree of heliocentric longitude from 315° to 335°. Using this miniature ephemeris, Challis began trial observations on July 29. The first sweep was made over the region where Adams expected the planet to lie, but Challis found that, using Airy's transit method, he could not record all the stars, which crossed the field in dense clusters. The next night he tested a second method in which the telescope was driven in declination, and right ascension was measured with a hand-operated hour circle. This system required Challis's two assistants, James Breen and John Holdsworth, to record the data. They were instructed to draw a line under the last observation if a cloud interrupted the sweep. Challis compared the two systems of observation (telescope fixed and moving) and found that each offered some advantages; he decided to observe the entire 10°-by-30° field using both methods, so as to avoid missing anything.

On August 4, the third night of observation, Challis recorded points of demarcation for the zones of 9′ breadth that were to be swept subsequently. The planned sweeps had to be postponed for a week because of moonlight and bad weather, and Challis took advantage of the interruption to report his progress to the Royal Observatory. Thinking, mistakenly, that Airy had already left for the Continent, he wrote to the chief assistant at Greenwich, the Reverend Robert Main: "I have undertaken to search for the supposed new planet more distant than *Uranus*. Already I have made trial of two different methods of observing. In one method, recommended by Mr. Airy . . . I met with a difficulty which I had anticipated . . . I adopted a second method." [10] Airy was still at Greenwich when this letter arrived. On August 10, he and his wife left for Wiesbaden, a well-known German spa. Two nights later, the sky cleared over Cambridge, and Challis resumed his search.

Challis decided to use August 12, the fourth night of observation, as a check point. He reswept, with the telescope fixed, the same zone he had examined on July 30 with the telescope moving. In the course of this sweep a cloud temporarily blocked off the field, and Challis's assistant dutifully drew a line under star No. 39, the last one recorded before the interruption. "Soon after August 12," Challis compared the observations made that night with those of July 30; he found that, as far as he carried the comparison, all the stars recorded in the first sweep were observed in the second one. He ended his check at what later proved to be a crucial stopping point: the line under star No. 39.

At about this time, three thousand miles away, an attempt was made to start another search for the hypothetical planet. Leverrier's second memoir on Uranus (the one submitted to the French Academy of Sciences on June 1, 1846) did not reach the United States until August. Sears Cook Walker, an astronomer on the staff of the United States Naval Observatory at Washington, D.C., read the paper and immediately suggested to his superintendent, Lieutenant Matthew Fontaine Maury, that a search be made for Leverrier's theoretical planet. Unfortunately, the Naval Observatory, like its counterpart at Greenwich, was overloaded with the routine work of positional astronomy, and Walker's search could not be fitted into the schedule until two months later.

In the meantime, while Challis continued his painstaking observations at Cambridge, Leverrier was putting the finishing touches to a third memoir on Uranus, which he presented to the Académie des Sciences on August 31, 1846. It was entitled "Sur la planète qui produit les anomalies observées dans le mouvement d'Uranus.—Détermination de sa masse, de son orbite et de sa position actuelle." The title was self explanatory; in his usual meticulous fashion, Leverrier had solved thirty-three equations of condition to yield the following elements for the disturbing planet:

Semimajor axis	36.154 a.u.
Sidereal period	217.387 years
Eccentricity	0.10761
Longitude of perihelion	284°45′
Mean longitude, January 1, 1847	318°47′
Mass	1/9300
True heliocentric longitude, January 1, 1847	326°32′
Distance from the Sun	33.06 a.u.

Leverrier pointed out that "this true longitude differs slightly from the value of 325° which resulted from my earlier researches. The [later] figure is based on additional, and more precise data; it places the planet about 5° to the East of the star δ Capricorni."

These coordinates left nothing to be desired in the way of completeness and precision, but Leverrier had learned by painful experience that a straightforward publication of the most accurate data could not overcome the inertia of the French observers. In an attempt to provoke some action, he listed all the advantages of looking for the planet immediately: "The opposition of the planet took place on August 19th. This is, therefore, a most favorable time to discover it. The advantage of its great angular distance from the Sun will diminish steadily." An important consideration was the planet's sizable disk: "at the moment of opposition, the new planet will subtend an angle of 3.3″. Assuming that the brilliance of the planet is adequate, a good refractor is perfectly capable of distinguishing such a disk from the spurious circles produced by various aberrations." On the basis of the surface reflectivity of the known planets, Leverrier continued, there was every reason to believe that the disk of the new planet could easily be distinguished from the stars:

> This is a most important point. If the object that we wish to discover can be confused, by its appearance, with the stars, then in order to distinguish between them it would be necessary to observe all the small stars in the pertinent region of the sky, and try to detect one among them with a proper motion relative to the others. This project will be long and difficult. But if, on the other hand, the planet's disk has a large enough diameter to preclude its being confused with the stars, if one can substitute for the rigorous determination of the positions of all luminous points, a simple inspection of their physical appearance, then the search will move along rapidly.

What effect did Leverrier's exhortation produce among his colleagues? None whatsoever. No shuttered domes slid open to reveal a Lerebours objective aimed at Capricorn, no French sidereal drive ticked over the quiet seconds of the 22nd hour of right ascension. Despite Arago's earlier enthusiasm, no official search was begun; ponderous indifference was the spirit of the day.

In England, Adams simultaneously faced the same barrier. Since leaving his results with Airy, he had revised his calculations twice more. On September 2, 1846, unaware that Airy was in Germany, Adams wrote to him at Greenwich, enclosing his sixth corrected solution to the problem of Uranus. Adams had reconsidered his dependence on Bode's law for the assumed mean orbital radius of the hypothetical planet. He now felt that his earlier investigation "could scarcely be considered satisfactory while based on any thing arbitrary; and . . . therefore determined to repeat the calculation, making a different hypothesis as to the mean distance."[11] He was wary of making too great a change, and settled on a mean distance about 1/30 part less than the earlier one. Using this new figure, he calculated the following elements:

Mean longitude, October 1, 1846	323°2′
Longitude of perihelion	299°11′
Eccentricity	0.12062
Mass	0.00015003

"The result," he wrote, "is very satisfactory, and appears to shew that, by still further diminishing the distance, the agreement between the theory and the later observations may be rendered complete, and the eccentricity reduced at the same time to a very small quantity." Adams was not confident that his improved results would have any effect on the Astronomer Royal; like Leverrier, he was becoming more and more dissatisfied with the inertia of his scientific countrymen. He closed his letter to Airy with the remark, "I have been thinking of drawing up a brief account of my investigation to present to the British Association."

Airy was not at home to take the hint. On the same day that Adams wrote to him at Greenwich, Challis wrote to Airy at Wiesbaden, reporting on his laborious progress:

> I have lost no opportunity of searching for the planet; and, the nights having been generally pretty good, I have taken a considerable number of observations: but I get over the ground very slowly, thinking it right to include all stars to 10–11 magnitude; and I find, that to scrutinise, thoroughly, in this way the proposed portion of the heavens, will require many more observations than I can take this year.[12]

Airy received Challis's note, but Adams's letter was not forwarded to him. Main, his assistant at Greenwich, answered it on September 5 by offering Adams more data on Uranus for the years 1844 and 1845. In retrospect, Main's letter seems naïve to the point of inanity, expressive of that intellectual emasculation which Airy encouraged in his subordinates. On September 7, Adams replied shortly to Main's offer: "I hope by tomorrow to have obtained approximate values of the inclination and longitude of the node." [13] These figures were to be included in the summary of his researches that he was now preparing for the British Association for the Advancement of Science.

As soon as he completed his paper, Adams hurried off to Southampton, where the meetings of September 1846 were in progress. When he arrived he was dismayed to find that Section A (Mathematical and Physical Science) had finished its sessions on the previous day; he was too late. The date was Tuesday, September 15. On September 10 John Herschel had given his valedictory address as president of the British Association. In his review of scientific events he mentioned that the past year had seen the discovery of a new minor planet (*Astraea,* discovered on December 8, 1845, after a fifteen-year search, by Karl Ludwig Hencke, a postmaster and amateur astronomer of Driessen, Prussia), and continued, "It has done more,—it has given us the probable prospect of another. We see it as Columbus saw America from the shores of Spain. Its movements have been felt, trembling along the far-reaching line of our analysis with a certainty hardly inferior to ocular demonstration." [14] John Adams must have heard of this remarkable statement with strong mixed emotions.

At least his feelings may have been tempered by the thought that Challis was actually searching for the planet. Leverrier had no such compensation, and halfway through September 1846 he finally ran out of patience. He had given French astronomers all the necessary data to make a great discovery, and they had persistently ignored his efforts. Leverrier was a patriotic man, and it was a measure of his frustration that he now forsook his scientific allegiance to his country and cast about for any astronomer who would search for the planet.

About a year before, Leverrier had received a publication from Johann Gottfried Galle (1812–1910), an assistant at the Berlin Observatory. Galle had submitted as his dissertation for the doctor's degree an analysis and reduction of the few surviving observations made by the pioneer Danish astronomer Olaus Roemer (1644–1710). The paper included a description of Roemer's instruments and techniques, the reduced positions of eighty-eight stars, compared with observations of them made by Bradley and Piazzi, and three observations of the Sun, the Moon, and the planets out to Saturn.[15] When the dissertation was published at Berlin in 1845, Galle sent a copy of it to Leverrier. Either correspondence was very leisurely in the mid-nineteenth century, or else Leverrier simply neglected to acknowledge his receipt of Galle's book. Now, a year later, he saw in it a chance to instigate a search for the hypothetical planet. On September 18, 1846, Leverrier wrote to Galle; his letter began with an effusive compliment and ended with an urgent request:

Sir,—I have read with much interest and attention the reductions of Roemer's observations which you have been kind enough to send me. The perfect clarity of your explanations, the complete rigor of the results which you present are on a par with those which we should expect from a most able astronomer. At some future time, Sir, I will ask your permission to review several points which interested me, in particular the observations of Mercury included in your paper. Right now I would like to find a persistent observer, who would be willing to devote some time to an examination of a part of the sky in which there may be a planet to discover. I have been led to this conclusion by the theory of Uranus. An abstract of my researches is going to appear in the *Astronomische Nachrichten.* I will then be able, Sir, to offer my excuses to you in writing, if I have not fulfilled my obligation to thank you for the interesting work which you sent me.

You will see, Sir, that I demonstrate that it is impossible to satisfy the observations of Uranus without introducing the action of a new Planet, thus far unknown; and, remarkably, there is only one single position in the ecliptic where this perturbing Planet can be located. Here are the elements of the orbit which I assign to this body: [the elements Leverrier sent to Galle were those published in his memoir of August 31, 1846; see page 111].

The actual position of this body shows that we are now, and will be for several months, in a favorable situation for the discovery.

Furthermore, the mass of the planet allows us to conclude that its apparent diameter is more than 3″ of arc. This disk is perfectly distinguishable, in a good glass, from the spurious star-diameters caused by various aberrations.

I am, Sir, your faithful servant,

U. J. Le Verrier

Will you convey to Mr. Encke, although I have not had the honor of meeting him, my compliments and deep respect.[16]

This letter reached Galle on September 23, and he immediately asked his superior, Johann Franz Encke (1791–1865), Director of the Berlin Observatory, for permission to search for the planet. (It must be mentioned here that Encke had the same attitude toward his assistant as Airy had toward his corps of "drudges"; Encke had deprived Galle of the credit for an indispensable contribution to an important paper on Saturn; he also tried to divert attention from Galle's discovery of Saturn's "crêpe ring" in 1838.) Encke was unreceptive to his assistant's proposal to look for the hypothetical planet. Galle persisted, and at last Encke reluctantly gave his permission. In the course of their discussion they were interrupted by Heinrich Louis d'Arrest (1822–1875), a young student-astronomer who had overheard Galle's request, and who begged to be allowed to participate in the search. Galle, who had won his point, felt that "it would have been unkind to refuse the wish of this zealous young astronomer."

That same night, September 23, 1846, Galle and d'Arrest opened the dome over the 9-inch Fraunhofer refractor that was the pride of the Berlin Observatory.[17] Galle had computed the geocentric coordinates of the planet from Leverrier's elements, and he trained the telescope at the point where the planet was predicted to lie: right ascension, $22^h\ 46^m$; declination, $-13°24'$. D'Arrest, excited but patient, watched quietly while Galle searched for the $3''$ disk that Leverrier had predicted. He could not find it, and d'Arrest suggested using a star map. Galle was skeptical; he had recently used Harding's chart of the same area and knew how inadequate it was. Nevertheless, he led d'Arrest to a cupboard containing all of the observatory's star maps. Sorting through a drawer of disordered charts, d'Arrest found a new map of the area Galle was searching. This chart, Hora XXI of the Berlin Academy's *Star Atlas,* had been printed before the end of 1845, but had not yet been distributed to other observatories.[18] Galle returned to the telescope; d'Arrest seated himself at a desk and began checking the stars on the map as Galle called out their appearances and positions. Only a few had been checked when Galle described a star of the eighth magnitude, right ascension $22^h\ 53^m\ 25^s.84$. D'Arrest exclaimed, "That star is not on the map!" It was, at last, the eighth planet.

CHAPTER 7

NEPTUNE

NEITHER Galle nor d'Arrest was certain they had found the new planet. The predicted disk was difficult to resolve, the motion almost imperceptible. Shortly after midnight, d'Arrest's excited report of the discovery brought Encke hurrying to the dome. The three astronomers tracked the object until it set at about 2:30 A.M., but they were still unsure of its identity. Not until the following night, September 24, 1846, were they able to make observations that definitely established the existence of the new planet. Its observed motion was about 3″ per hour, retrograde (Leverrier had predicted a daily motion of −68″.7) ; its diameter, measured with illuminated micrometer wires at a power of 320, was determined as 2.9″ by Encke and 2.7″ by Galle. Encke made a second measurement of the diameter, obtaining 3.2″; he was amazed by the coincidence of this result with the estimate of 3.3″ calculated by Leverrier.

The accuracy of the predicted coordinates was even more remarkable. Leverrier's elements gave the geocentric longitude of the planet on September 23.5, 1846, as 324° 58′; the geocentric longitude observed by Galle at September 23.50001 ($12^h\ 0^m\ 14^s.6$) was 325° 52′ 45″. The difference was about 55′ of arc—less than a single degree. On September 25 Galle triumphantly wrote to Leverrier:

The planet whose position you have pointed out *actually exists*. The same day that I received your letter, I found a star of the eighth magnitude which was not shown on the excellent chart (drawn by Dr. Bremiker), Hora XXI of the series of celestial maps published by the Royal Academy of Berlin. The observations made the following day determined that this was the sought-for planet.[1]

Galle sent Leverrier the coordinates of the reference stars (*Bessel,* Zone 119, $21^h\ 50^m\ 31^s$, $-13^\circ\ 30'\ 7''.9$ and *Piazzi,* Hora XXI, No. 344) that had been used to determine the planet's position and motion. He also suggested that the new planet might appropriately be named *Janus.*

During the following week, letters announcing the discovery left Berlin with increasing frequency. On September 26, Encke sent a report to Heinrich Christian Schumacher, editor of the *Astronomische Nachrichten,* and also wrote to Peter Andreas Hansen at Gotha. When Encke's letter reached Gotha on September 29, George Airy and his family had just arrived there for a three-day visit with Hansen. The Astronomer Royal thus learned of the discovery several days before the news reached England. He did not write to anyone about it for more than two weeks, and his first letters showed that the announcement had been for him the distant warning of a violent and inevitable storm.

On September 28 Encke wrote a gracious letter of congratulation to Leverrier:

Allow me, Sir, to congratulate you most sincerely on the brilliant discovery with which you have enriched astronomy. Your name will be forever linked with the most outstanding conceivable proof of the validity of universal gravitation, and I believe that these few words sum up all that the ambition of a scientist can wish for. It would be superfluous to add anything more.[2]

Encke was sincere in all but the last line; he went on to add another page of florid prose to his letter. Schumacher's message was written on the same day:

> Although you will have already heard from M. Encke, that your planet has been found in almost precisely the place and circumstances which you predicted (even to having a diameter of 3″), I cannot resist my heartfelt impulse to send you immediately my most sincere congratulations on your brilliant discovery. It is the most noble triumph of theory which I know of.[3]

While letters of congratulation were converging on Paris, Challis, unaware that the discovery had been made, was still plodding through his ponderous survey without much hope of completing it until 1847. On September 29, six days after the discovery, he first saw Leverrier's memoir of August 31 in which the French astronomer had emphasized (as we have seen) that the planet should be identifiable by its disk. Challis immediately accepted from Leverrier the suggestion he had ignored when Adams proposed it late in July. That same night he "swept a considerable breadth in Declination, between the limits of Right Ascension marked out by M. Le Verrier, and . . . paid particular attention to the physical appearance of the brightest stars." [4] In the course of this sweep, during which he observed about 300 stars, Challis hesitated over one object, and then directed his assistant to note, "It seems to have a disk." [5] He did not take the time to check it at a higher magnification; he later said that he intended to do so on the following night.

The Reverend William Towler Kingsley, a fellow and tutor of Sidney Sussex College, later provided an illuminating sidelight on this incident. Kingsley was dining with Challis at Trinity College one evening when Challis mentioned, in the course of describing his observations, that he had noted one star which seemed to have a disk. "Kingsley asked him, 'Would it not be worth while to look at it again with a higher power?' Challis replied, 'Yes, if you will come up with me when dinner is over we will have a look at it.' " [6] The sky was clear when the two men arrived at Challis's lodgings in the observatory. They had intended to go right up to the dome, but Mrs. Challis insisted on giving them tea before they began their observations. The delay was crucial: by the time they had finished their tea the sky had clouded over and observations were impossible.

There is not, unfortunately, any record of the exact date of this occurrence. Although the circumstantial evidence is conflicting, Kingsley's account suggests not only that "Challis omitted to follow up a valuable clue," but also that he "missed the great opportunity of his life either in the second week of September or on the last day of the month."

Challis did not have a chance to make an independent verification of his suspicions about the star with a disk. The next night, September 30, "the moon being in the way . . . the observations could not be repeated." [7] On October 1, the following letter was published in the London *Times:*

Sir,—I have just received a letter from Dr. Brünnow, of the Royal Observatory at Berlin, giving the very important information that Le Verrier's planet was found by Mr. Galle on the night of September 23rd . . . This discovery may be justly considered one of the greatest triumphs of theoretical astronomy.

I remain, Sir, your most obedient servant,
J. R. Hind

Mr. Bishop's Observatory, Regent's-park
Wednesday night

P.S. The planet has been observed here this evening; notwithstanding the moonlight and hazy sky, it appears bright, and with a power of 320 I can see the disk.

John Russell Hind (1823–1895) had been an assistant at the Greenwich Observatory from 1840 to 1844. At the time this letter was written he was the director of George Bishop's private observatory in Regent's Park, London. The letter that he received was written by Franz Friedrich Ernst Brünnow, a young German astronomer, on September 25. The largest telescope at Hind's disposal was an equatorially mounted refractor with a focal length of nearly 11 feet and an aperture of 7 inches. On September 30, the same night that Challis made no observations with the 11.75-inch Northumberland telescope because of moonlight (he did not mention any other deterrents), Hind observed the new planet and resolved its disk with his 7-inch refractor.

On October 1, when Challis learned of the discovery, he decided to analyze the 3150 star positions he had recorded since July 29. He began with the comparison of the observations made on July 30 and August 12, which, as we have seen, he had previously terminated at star No. 39, "probably from the . . . circumstance that a line was there drawn in the memorandum-book in consequence of the interruption of the observations by a cloud." [8] Now, on continuing the check, Challis was chagrined to find that "No. 49, a star of the 8th magnitude in the series of August 12, *was wanting in the series of July 30* . . . this was the planet." Furthermore, "After ascertaining the place of the planet on August 12, [he] readily inferred that it was also among the reference stars taken on August 4." Challis had thus recorded the planet's position twice in the first four days of observing. It took only a moment to verify that the object whose disk he had noticed on September 29 was also the new planet. Challis was mortified; he wrote to Airy, "after four days of observing, the planet was in my grasp, if only I had examined or mapped the observations." [9]

While Challis repented his indecision, Galle received Leverrier's thanks for making the crucial observation. Leverrier wrote on October 1:

Sir,—I thank you cordially for the alacrity with which you applied my instructions to your observations of September 23rd and 24th. We are thereby, thanks to you, definitely in possession of a new world. The pleasure I have felt, because you found it less than one degree from the position which I had given, is a little sullied by your writing so soon; we could have reached four months ago the result we have just achieved. I will transmit your letter to the Academy of Sciences next Monday. Allow me to hope that we will continue to carry on a correspondence which begins under such auspicious circumstances.

U. J. Le Verrier

The Bureau of Longitudes here has decided on *Neptune*. The symbol is to be a *trident*. The name Janus would imply that this is the last planet of the solar system, which we have no reason at all to believe.[10]

On the same day he wrote to Galle, Leverrier notified Airy and Friedrich Georg Wilhelm Struve, director of the Pulkovo Observatory, of the choice of the name Neptune. This was the beginning of the confusion that attended the naming of the new planet.

Galle originally suggested the name Janus because he had learned that Janus was a predecessor of Saturn. Leverrier rejected the name on the mistaken assumption that Janus, the two-faced Roman deity of gates and doors, hence of all beginnings, was also the god of boundaries. This error was corrected with great emphasis by James Pillans, professor of Latin at Edinburgh University, who pointed out that Leverrier was assigning to Janus the function of Terminus, the god of limits. Pillans was the only supporter of Galle's proposal; his colorful defense of the name Janus enlivened the formula-strewn pages of the *Astronomische Nachrichten,* but his arguments failed to influence any astronomers.

The name ostensibly proposed by the French Bureau of Longitudes, and communicated by Leverrier, was adopted by Gauss, Encke, Struve, Airy, and Herschel, and most other astronomers soon followed their lead. There was one peculiar circumstance about all this: it is certain that the Bureau of Longitude had nothing to do with the selection of the name Neptune. This department, roughly equivalent to the British Nautical Almanac Office, had never before named a planet; it was outside their competence to do so. Their minutes show, first, that they had not considered assigning a name to the new planet before October 1, when Leverrier wrote to Airy, Struve, and Galle, and, second, that the Bureau explicitly repudiated Leverrier's statement at a subsequent meeting. The name Neptune was certainly Leverrier's own suggestion. Since he had the discoverer's right to name the planet, why did he attribute the name he chose to a government office that claimed to know nothing about it?

There were two possible explanations. The first was a desire for the weight of authority behind Leverrier's preference. Although Galle had modestly assigned all the credit for the discovery to Leverrier, the latter may have felt that his glory would be diminished if Galle's suggested name for the planet was adopted. A bogus official sanction was one way to discourage Galle at the outset. This is not a flattering supposition, but it is not at all improbable in the light of Leverrier's subsequent actions. The second possibility is that Leverrier made a sincere but unwarranted extrapolation. On October 26, 1846, he announced his intention to publish a complete memoir on Uranus in the *Connaissance des temps,* which was issued by the Bureau of Longitude. He may have mentioned the name Neptune to one or more members of the Bureau, and, in his haste to answer Schumacher's request of September 29 for the new planet's name,[11] expanded the approval of a few individuals into an official sanction. Whatever the reason, his own satisfaction with the name Neptune lasted only a few days.

Sometime between Thursday, October 1, and Monday, October 5, 1846, Leverrier decided that he wanted the planet named after himself. He implored Arago, who had already accepted the name Neptune, "as a friend and as a countryman, to adopt the name Leverrier." Arago finally agreed, but only on the condition that the name Herschel, which he had long favored, should replace Uranus. When Leverrier's combined memoir on the seventh planet appeared a few months later, it was entitled, "Recherches sur le mouvement de la planète Herschel (dite Uranus)." Although the planet was invariably referred to as Uranus in the text, the author inserted the following note: "In my future researches, I shall consider it my strict duty to eliminate the name Uranus completely, and to call the planet only by the name *Herschel.* I regret that the printing of this memoir has advanced to where it has prevented me from carrying out a resolution which I shall observe religiously in the future."

At the meeting of the French Academy of Sciences held on Monday, October 5, Arago presented Galle's letter describing the discovery, and then announced that, "having received from M. Le Verrier a most flattering invitation, the privilege of naming the new planet, [he] made up his mind to call it by the name of the man who so learnedly discovered it, to name it *LeVerrier.*" [12] Arago was well aware of the controversial implications of this decision, and he defended it in an impassioned speech:

> I did not believe that I should be deterred by a few baseless objections. Why indeed! Comets are named after the astronomers who discovered them, or who traced out their orbits; should the same honor be refused to the *discoverers* of planets! We have, as is just, Halley's comet, Encke's comet; is it right that we should have comets named for Gambart or Biela, de Vico, Faye, and others, and that the name of the man who, by an admirable and unprecedented method, has demonstrated the existence of a new planet, specified its place and size, should not be written in the heavens! ! ! No, no! It offends reason and the most common principles of justice.
>
> Is there apprehension about the few reforms which my resolution would seem to entail? Well, they are no obstacles: Herschel will dethrone Uranus; the name of Olbers will replace that of Juno, and so on; It is never too late to tear off the swaddling clothes of habit!
>
> I pledge myself never to call the new planet by any other name than *LeVerrier's Planet.* I believe I will thus give an irrefutable proof of my love for science, and [also] follow the inspiration of a legitimate patriotism.[13]

On October 6 Arago reported to Schumacher his adoption of the name LeVerrier for the new planet, and the notice was printed in the *Astronomische Nachrichten* for November 5.[14] It was received very unfavorably, as were the private announcements that Leverrier sent to various astronomers. Leverrier's note of October 6 to Encke (who, as we have seen, had already adopted the name Neptune) was coy and misleading: "I requested my illustrious friend M. Arago to undertake the duty of naming the planet. I was a little confused by the choice he made in the intimate circle of the Academy." Encke received this message almost simultaneously with one written by Gauss on October 7: "I find the name Neptune, chosen by *Leverrier,* perfectly appropriate; a trident might be acceptable for the symbol, if it would not be improper for the author to hold it." On October 22 Encke stated in a letter to the *Astronomische Nachrichten* that he, Gauss, Herschel, and Struve were satisfied with Neptune because it was mythological, appropriate, and had (they believed) been approved by the discoverer. They were not willing to relinquish it for a title that glorified one man at the expense of an inoffensive and long-established tradition. After this time, the name Neptune prevailed on the Continent (although Arago did not capitulate for years), but in England one more attempt was made to name the planet. Most European astronomers were surprised to learn that anyone in England assumed the right to do so.

The first public notice of John Couch Adams's work was a letter written by John Herschel which appeared in the London *Athenaeum* on October 3, 1846. After quoting the dramatic prophecy he had made at the meeting of the British Association only three weeks before, Herschel wrote:

The remarkable calculations of M. Le Verrier—which have pointed out, as now appears, nearly the true situation of the new planet, by resolving the inverse problem of the perturbations—if uncorroborated by repetition of the numerical calculations by another hand, or by independent investigation from another quarter, would hardly justify so strong an assurance as that conveyed by my expression above alluded to. But it was known to me, at that time, (I will take the liberty to cite the Astronomer-Royal as my authority) that a similar investigation had been independently entered into, and a conclusion as to the situation of the new planet very nearly coincident with M. Le Verrier's arrived at (in entire ignorance of his conclusions), by a young Cambridge mathematician, Mr. Adams;—who will, I hope, pardon this mention of his name (the matter being one of great historical moment),—and who will, doubtless, in his own good time and manner, place his calculations before the public.[15]

This notice caused a ripple of excitement in England, but for the time being it went unnoticed on the Continent. French astronomers first learned of the British search for Neptune when Arago received a letter written by Challis on October 5. Considering the circumstances, it was a most peculiar message. Challis wrote that, in the course of constructing a star map, he had been looking for the new planet since July 29, but without success until he read Leverrier's third memoir on the 29 of September. That night he had noticed one star that seemed to have a disk, and, as he now informed Arago, it was the new planet. He pointed out that his observation "confirmed, in a remarkable manner, the accuracy of the conclusion which M. Le Verrier arrived at from theoretical considerations, that with a good telescope the planet could be distinguished by its physical appearance." [16] Challis did not explain why he had been looking for the planet since July 29 or whose data he had been using. He made no reference to Adams or to any independent prediction of Neptune's position. The French astronomers reasonably assumed that Challis had used Leverrier's data; when he later denied it they were angry and skeptical.

George Airy was already spreading oil in preparation for the coming storm. On October 14, he sent Leverrier

sincere congratulations on this successful termination to your vast and skilfully directed labours. I was exceedingly struck with the completeness of your investigations. May you enjoy the honours which await you, and may you undertake other work with the same skill and the same success and receive from all the enjoyment which you merit.

I do not know whether you are aware that collateral researches had been going on in England and that they led to precisely the same results as yours. I think it probable that I shall be called on to give an account of these. If in this I shall give praise to others, I beg that you will not consider it as at all interfering with my acknowledgement of your claims. You are to be recognized beyond doubt as the real predictor of the planet's place. I may add that the English investigations, as I believe, were not quite so extensive as yours. They were known to me earlier than yours.[17]

On the same day he wrote to Leverrier, Airy replied to a letter from Challis, dated October 12, in which the latter described his observations. The Astronomer Royal did not hesitate to change his opinions to suit their destination; he told Challis, "Heartily do I wish that you had picked up the planet, I mean in the eyes of the public, because in my eyes you have done so. But these misses are sometimes nearly unavoidable." In his letter Challis had suggested naming the new planet Oceanus. Airy (who had already given his approval of the name Neptune) missed no tricks; his reply continued, "I like your name *Oceanus*." He concluded by asking Challis for permission to publish the previous correspondence between them on the subject of the planet.

Airy made the same request of John Adams on the same day:

It appears to me proper that I should write and communicate to the Royal Astronomical Society an account of what I know of the English transactions [relating to the new planet]. My reasons, as I have explained to Professor Challis, are: (1) that it is illustrative of the history of Science, (2) that it would do justice to England, (3) that it would do justice to individuals, (4) I could do it because I know nearly all the history and yet have taken no part in the theory or the observations. It would be wrong for me to compromise any one and therefore I ask no advice about this. All that I ask is: Will you permit me to publish your correspondence with me on this subject, or extracts from the correspondence taken at my discretion?[18]

There was at least one point on which Airy's claim to "know nearly all the history" could be disputed; he sent this letter to the "Rev. W. J. Adams." Despite the incorrect address, Adams received the note, and he replied to it on the next day, October 15. He agreed to the publication of his correspondence with Airy, and also sent a partial set of elements for Neptune, based on the positions recorded by Challis on August 4 and 12, and observations made after the planet's discovery, up to October 13. Adams's postscript chided Airy for his lapse of memory: "I may mention that I am not yet in Orders."

By this time Leverrier was aware that his claims were being contested by a young Englishman (of whom he had never heard) named Adams. Though the situation was not yet very clear, Leverrier was angry and upset. On October 16 he wrote to Airy and asked some pointed questions about the Astronomer Royal's letter of June 26, 1846. If, as Airy now claimed, he had known then about the remarkable coincidence of Adams's and Leverrier's results, why had he not mentioned it? For that matter, "Why has Mr. Adams kept silent for four months? Why had he not spoken since June if he had good reasons to give? Why did he wait until the planet was seen in the telescope?" And if Challis had been using Adams's data, why had he mentioned only Leverrier's name in his letter of October 5? Leverrier could not have known how potentially embarrassing these questions were for Airy; he asked the Astronomer Royal to intercede for him and defend his rights.

Airy replied on October 18. After expressing his regret that Leverrier had been so troubled by Herschel's letter to the *Athenaeum,* he added, "I am confident that you will find that no real injustice is done to you; and I hope that you will receive this expression the more readily from me, because I have not hesitated to express to others, as well as to yourself, very strong feeling upon the extraordinary merit of your proceedings in the matter." Airy's last-minute attempt to avert an open conflict was too late. On October 17 Challis's detailed account of the British investigation was published in the *Athenaeum*. His report included an outline history of Adams's work, beginning with the "resolution" of 1843, and described the Cambridge observations up to August 12, 1846. Challis even suggested the name Oceanus for the new planet. When this publication reached France the storm broke.

The center of turbulence was the chamber of the Academy of Sciences, in which occurred, on October 19, 1846, a session unlike anything those precincts had witnessed for decades. A leading Parisian journal reported of it, "What a meeting! Are we in the Academy of Sciences or in the Chamber of Deputies? Is it a question of the existence of M. Guizot's Ministry or M. Leverrier's planet?" [19] If emotional fervor was the criterion, it would have been hard to tell. François Arago began the session with an "Examination of the critical remarks and the questions of priority which M. Le Verrier's discovery has provoked." He first established that "the planet LeVerrier [was] the first one whose existence and position were predicted by theory," and asserted that the planet had never been observed before Leverrier published his researches.

Arago's manner became more emotional when he took up "the question of priority, raised a few days ago in England, by Sir John Herschel, Mr. Airy, director of the Greenwich Observatory, and Mr. Challis, director of the Cambridge Observatory." He presented translations of various letters written by these men, and then made "a detailed refutation" of their claims. The first two items were Herschel's letter of October 1 to the *Athenaeum* and Airy's October 14 letter to Leverrier. Herschel's statement that his prediction was justified only by the coincidence of Adams's and Leverrier's results was decried as "hardly civil, in complete contrast with the usual courtesy and reserve of Mr. Herschel." Airy was (more justly than his attacker knew) discredited because of his preoccupation with the radius vector. Arago quoted part of the query Airy had sent to Leverrier, and then abruptly broke off: "It would be superfluous to carry the citation any farther. That which I have quoted shows that Mr. Airy, one of our most learned contemporaries in the theory of celestial mechanics, would not believe, though in possession of Mr. Adams's work, that the action of an exterior planet could account for the anomalous variations in the radius vector of Uranus."

"These two articles," continued Arago, "can only be regarded as skirmishes. We will join full battle in giving the translation of a letter by Mr. Challis, director of the Cambridge Observatory, addressed to the English journal *The Athenaeum.*" [20] This was Challis's detailed account described above; it was translated in its entirety, and Arago proceeded to dismember it line by line. Challis's position was indefensible. The disparity between his letter of October 5 in which he attributed his discovery to Leverrier's memoir and his report to the *Athenaeum* in which he mentioned only Adams's work made him appear either dishonest or demented. Arago commented on Challis's letters with condescending sarcasm: "I will not seek to reconcile the two versions. I will leave it to Mr. Challis to explain how the name of Adams, which did not appear at all in his first message, became so prominent in the second."

Arago summed up his argument with the statement that there had been "neither mention of any publication of Adams's work, nor any example of it shown. This circumstance alone suffices to end the debate. There exists only one rational and just way to write the history of science: it is to rely exclusively upon *publications* bearing a definite date; abandoning that, all is confusion and obscurity." [21] Encouraged by the outbursts of approval that greeted his remarks, Arago delivered an impassioned closing oration:

> Mr. Challis so exaggerates the merit of Mr. Adams's *clandestine* work, that he assigns to the young Cambridge geometer the right to name the new heavenly body. This claim will not be accepted. The public owes no debt to him who has taught it nothing, who has rendered it no service. What! M. Le Verrier has made his researches available to the entire scientific world; following the formulas of our learned compatriot, everyone was able to see the new planet dawn, rapidly become brighter, and before long appear in its full brilliance; and today we are called upon to share this glory, acquired so diligently and legitimately, with a young man who has communicated nothing to the public, and whose calculations, more or less incomplete, are with only two exceptions totally unknown to the Observatories of Europe! No, no! The friends of science will not permit the perpetration of such a flagrant injustice! The journals, the Letters which I have received from several English scientists prove to me that in that country too, the eminently respectable rights of our compatriot will find zealous defenders.
>
> Mr. Adams has no right to figure in the history of the discovery of the planet Le Verrier, neither by a detailed citation, nor by the slightest allusion.
>
> In the eyes of every impartial man, this discovery will remain one of the most magnificent triumphs of astronomical theory, one of the glories of the Academy, and one of our country's noblest titles to the gratitude and admiration of posterity.[22]

The French press began where this speech ended. On October 21, *L'Univers* attacked England violently, denouncing "an odious national jealousy which appears to be determined not to let us have the last word." *Le*

National of the same date began a tirade that was to last for a month, and accused the English of "treating France as a *stupid nation,* M. Arago as a *Humbug,* and our own writings as *discredited* articles, everything crowned by the glorious refrain: 'Adams and England for ever.'" On November 7, *L'Illustration* published five brutally satirical caricatures of Adams. By that time the attacks in the French journals were so unrelated to the facts that both Arago and Leverrier publicly disavowed their sympathy with the press.

Airy was still trying to reconcile the two sides. He wrote to Leverrier on October 21, apologizing for Challis's letter to the *Athenaeum,* and explained that it was only meant to acquaint the British public with Adams's work.[23] To his dismay, this letter was published and savagely attacked in the French papers. His next letter to Leverrier, dated October 26, was written primarily to extricate himself from the controversy. He pointed out that nothing concerning the new planet had been published under his name, and also expressed astonishment at Leverrier's venomous opinion of Challis, especially since Challis had never questioned Leverrier's priority of discovery. Despite Airy's intercession, Challis came in for more than his share of French antagonism. One Academician searched out some of Challis's old papers and after subjecting them to microscopic scrutiny published a vindictive memoir, "On the serious errors committed by a foreign mathematician." [24] The resurrected papers were more than thirteen years old, and the errors were trivial, but the attempt to discredit Challis must have cost him considerable pain.

By the end of October, both Airy and Challis were under heavy fire in Britain, and it was clear that they would have to make some public explanation of their actions. On November 13 Airy read to the Royal Astronomical Society his "Account of some circumstances historically connected with the discovery of the Planet exterior to Uranus." [25] Challis, who had also been asked to make a report, was understandably nervous about his position, but, though he wrote to Airy on October 3, "I am in difficulties about this report and should be glad to see some means of getting out of it," he found no way out. His paper was entitled, "Account of Observations at the Cambridge Observatory for detecting the Planet exterior to Uranus." [26] At the same meeting Adams presented "An Explanation of the observed Irregularities in the Motion of *Uranus,* on the Hypothesis of Disturbance caused by a more distant Planet; with a Determination of the Mass, Orbit, and Position of the disturbing Body." [27]

What were the reactions to "the three most remarkable communications . . . which the Society can ever expect to receive in one night"? Adams's paper was applauded as a masterpiece of succinct analysis. Challis made a very bad showing. His attackers wanted to know, why had he not examined his data earlier? For that matter, why had he not looked for the planet in September 1845, when Adams had finished his first complete set of elements? The answer to both questions was the same: lack of confidence. It was true that, in 1845, by the time Adams had finished his calculations the planet was well past opposition, and hence in an unfavorable position for detection. Nevertheless, Challis admitted that he had been deterred chiefly because it seemed "so novel a thing to undertake observations in reliance upon merely theoretical deductions, and that while much labour was certain, success appeared very doubtful." In 1846 Challis had not mapped his observations, "partly because I thought the probability of discovery was small till a much larger portion of the heavens was scrutinised, but chiefly because I was making a grand effort to reduce the vast number of comet observations which I have accumulated; and this occupied the whole of my time when I was not engaged in observing." In evaluating what might seem to be a makeshift excuse, we should consider the comment made by Ralph Allen Sampson, a historian of the Royal Astronomical Society: "To do ['Challis's pitiful story'] justice, it is candid. No one would dream of doubting its veracity, for what could induce any man to produce a tale of that complexion?" [28] Through his own lack of initiative and his submission to Airy's intellectual domination, Challis had lost his chance to make a great discovery and forfeited the respect of his colleagues.

Airy was not exonerated by his "Account." In his autobiography he wrote that during this period he was "abused most savagely both by English and French." Outstanding among the critics were the Scottish physicist Sir David Brewster and the terrible-tempered Sir James South, each of whom wrote long unsparing attacks of the Astronomer Royal. Airy was informed of the widespread adverse opinion of his actions by one of his close friends, the geologist Adam Sedgwick. On December 3, 1846, Sedgwick wrote to Airy:

> You were accused, not only of unreasonable incredulity and apathy toward Adams of St. John's, but also of having (as was said) "snubbed him from the first" and so acting on a timid person prevented him from reaping the honors of a great discovery . . .
>
> As to apathy and incredulity misapplied, I think the facts speak so loudly that my dull ears cannot help hearing them. Your own statement is clearly and honestly written and certainly is not written in the spirit of self-praise. Had the results communicated to you and Challis been sent to Berlin, I am told, they came *so near the mark* that to a certainty the new planet would have been made out in a very few weeks, perhaps a very few days, and the whole business settled in 1845—Adams the sole, unadvised, unassisted discoverer. Is this true?

Do Adams's results come so near the mark as to justify such language? If so, I must myself chime in with the pack of grumblers. To say the very least of it, a grand occasion has been thrown away.[29]

This much-moderated version of the prevailing British attitude had almost no effect on Airy. Criticism of him

simply realized the old fable of the viper and the file. Attacks which would have agonized Flamsteed's every nerve, and have called forth full and dignified rejoinders from Maskelyne, were absolutely and entirely disregarded by Airy. He had done his duty, and in his own estimation . . . had done it well. He was perfectly satisfied with himself, and what other people thought or said about him influenced him no more than the opinions of the inhabitants of Saturn.[30]

During the last months of 1846 many angry inhabitants of France were still trying to break through Airy's wall of self-esteem. Part of their hostility was an expression of Anglo-French political rivalry; it is pertinent to ask to what extent political affairs influenced the attitudes of scientists toward the Adams-Leverrier controversy. The crucial international issue at this time was the conflict over the Spanish marriages. On October 10, 1846, Isabella II, the reigning queen of Spain, was married to the Duke of Cadiz; at the same ceremony the queen's sister, the Infanta Luisa, became the wife of the Duke of Montpensier, the fifth son of Louis Philippe of France. Though in some respects provoked by the reckless actions of Lord Palmerston, the British foreign secretary, these marriages and the political alliance they implied were directly opposed to the aims and previous agreements of the British foreign office. The British had been outwitted by Louis Philippe, and they were not happy about it:

Louis Philippe and M. Guizot had attained their object, but they had sacrificed "the cordial understanding." In England, the authors of the Spanish marriages stood condemned by the Court, by the Cabinet, and by public opinion. There was no thought of an appeal to arms, but it was felt that in the future the relations of the two governments could no longer be carried on in the same spirit of close and friendly intimacy.[31]

The Spanish marriages were celebrated just before the discovery of Neptune became a public issue, and at least one French journal tried to relate the two incidents. On October 21, *L'Univers* complained that "for want of pride, and out of sordid envy, [the English] would like to make us pay the wedding costs of the Spanish marriages with M. Leverrier and his planet." If chauvinism had influenced the opinions of many British scientists, it is unlikely that they would have defended Leverrier after Adams's work became known. This, however, is exactly what they chose to do. Not only did most British astronomers support Leverrier's claims, but on November 30, 1846, the Royal Society of London, which had known of Adams's work for several weeks, awarded its highest scientific honor, the Copley medal, to Leverrier for his brilliant discovery of the new planet. The members of the Royal Astronomical Society were too close to the controversy to make a similar decision; after heated debate they elected to award no medal for the year 1846: "it seems to have been thought by several that an award to M. Le Verrier, unaccompanied by another to Mr. Adams, would be drawing a greater distinction between the two than fairly represents the proper inference from facts, and would be an injustice to the latter."[32]

By the end of 1846 Neptune had been observed by most astronomers in Europe and America. One important discovery was made by William Lassell. On October 10, 1846, less than three weeks after Galle found the planet, Lassell thought he saw a satellite of Neptune. He could not confirm his suspicion until July 1847; the satellite was named Triton. Sears Cook Walker, the American astronomer who had tried to start a search for Neptune at the United States Naval Observatory, suspected that because of its very slow motion the planet might have been recorded as a star prior to its discovery. In January 1847 Walker began to look for "ancient" observations and quickly eliminated all star catalogues compiled earlier than Lalande's *Histoire céleste française*. He found that Lalande had observed in the vicinity of Neptune on May 8 and 10, 1795, and that an eighth-magnitude star recorded by Lalande on May 10 had been within 2′ of the position of Neptune at that date. The star was not shown in Bessel's later zone catalogues of the same region. Walker made this discovery on a cloudy night, February 2, 1847, and he asked to have the star searched for on the next clear evening, expecting that it would be missing. On February 4, Professor Joseph Hubbard swept the region and confirmed Walker's expectation; the star was not there. Using Lalande's probable "ancient" observation and many modern ones, Walker calculated new elements for Neptune. His results were not published in Europe until May 20, 1847.[33]

In the meantime, Schumacher's assistant, Dr. Adolf Cornelius Petersen (1804–1854), independently discovered Lalande's observation by a slightly different method. His report was published in the *Astronomische Nachrichten* on April 3, 1847.[34] Neither Petersen nor Walker was sure of the observation because Lalande had followed it with (:), his symbol for a doubtful position. Félix Mauvais tried to resolve the difficulty by examining Lalande's original manuscript. His research provided a striking confirmation of Walker and Petersen. Mauvais found that Lalande had observed the star *twice,* on May 8 and 10, but because the two positions did not agree he had deleted the first observation from the printed catalogue and marked the second doubtful. The difference between the two positions was found by Walker and Mauvais to be precisely equal to the movement of Neptune in two days.

Walker's elements for Neptune caused a sensation in Europe and America, since they bore almost no resemblance to those calculated by Adams or Leverrier. Walker showed, first, that Bode's law, which predicted a mean orbital radius of 39.2 a.u. for the eighth planet, broke down completely in the case of Neptune, and, second, that the only point on which the theoretical and observed elements did not differ markedly was the true longitude. The three sets of elements compared as follows:

	Adams	*Leverrier*	*Walker*
Semimajor axis (a. u.)	37.25	36.15	30.25
Eccentricity	0.12062	0.10761	0.00884
Orbital period (years)	227.3	217.387	166.381
Mass (Sun = 1.0)	1/6666	1/9300	1/15000
Longitude of perihelion	299°11′	284°45′	0°12′25″
True longitude, January 1, 1847	329°57′	326°32′	328°7′34″

The remarkable differences prompted Professor Benjamin Peirce (1809–1880) of Harvard University to declare

that the planet Neptune is not the planet to which geometrical analysis has directed the telescope; that its orbit is not contained within the limits of space which have been explored by geometers searching for the source of the disturbances of Uranus; and that its discovery by Galle must be regarded as a happy accident.[35]

Although supported by mathematical arguments, Peirce's conclusion was drastically overstated. As W. M. Smart, the first John Couch Adams Astronomer of Cambridge University, has pointed out, the discovery was certainly not an accident; the elements computed by Adams and Leverrier were accurate enough to accomplish their purpose. For the period 1801–1843, "the perturbations in longitude on Uranus are dependent on the mass of Neptune, the distance between the two planets and the difference in their heliocentric longitudes." The perturbations would not be much different if the heliocentric distance of Neptune was increased and the diminution of its attractive force on Uranus compensated for by increasing its mass:

> In other words the assumption of any value of the semi-major axis of the unknown planet's orbit, not differing too wildly from the true value, would lead to values of Neptune's heliocentric longitudes, at any time between 1801 and 1843, within a few degrees of the true values, thus providing adequate information for the telescopic detection of the unknown planet.[36]

Few Europeans accepted Peirce's conclusions, and Leverrier was showered with honors for the discovery of Neptune. On October 5, 1846, the king of France had named him an officer of the Legion of Honor, and soon afterward he was designated the first occupant of a chair of celestial mechanics established by royal ordinance in the Faculty of Science at Paris. The Royal Society of London's Copley medal was one of many awards conferred on Leverrier by groups outside of France. The Royal Society of Göttingen elected him a foreign associate; the Imperial Academy of Sciences of Saint Petersburg sent him their congratulations and set aside their next vacancy for him. The King of Denmark made Leverrier a commander of the Royal Order of Dannebroga, and the Grand Duke of Tuscany presented him with a new edition of Galileo's works and the collected memoirs of the Accademia del Cimento.

Although years passed before the disparity between Adams's and Leverrier's honors was made up, by early 1847 rational opinions were beginning to prevail about the relative merits of their accomplishments. John Herschel and the French astronomer Jean Baptiste Biot (1774–1862) were outstanding among the peacemakers. In an article published in February 1847, Biot wrote:

> I do not speak here in that narrow spirit of geographical selfishness so wrongly called patriotism. The minds dedicated to the pursuit of science belong, in my eyes, to a common intellectual nation, which includes all the degrees of elevation of the pole. I only see a talented young man who has been served badly by circumstances this one time, and whom one must applaud in spite of it. I say to him: The laurel which you merited first, another merited also, and has carried it away before you had the boldness to seize it. The discovery belongs to him who proclaimed it publicly, and announced to everyone what you were meanwhile keeping secret . . . But, in your own mind, you know that the new planet was known in theory to you before any other person . . . "Persevere."[37]

A few months after this was published Adams and Leverrier met for the first time, at the June 1847 meetings of the British Association for the Advancement of Science, and after the meetings John Herschel invited them both to his home at Collingwood. Neither Adams nor Leverrier had become embittered by the public controversy; to Herschel's delight they were immediately taken with one another. One guest remembered "being a witness of the first meeting between Professor Adams and Le Verrier and being charmed, like everyone else, at the cordial handclasp that was exchanged." The Reverend W. T. Kingsley (the man who had the crucial cup of tea with Challis) later wrote: "One of the pleasantest things in my life to look back to is a walk I had with Adams and Leverrier, the two men evidently admiring one another and perfectly free from jealousy."

Adams and Leverrier remained friends for their entire lives. The codiscoverers of Neptune—for that is how they ultimately came to be known—both made many other important contributions to astronomy. Leverrier died in 1877, Adams in 1892; in retrospect, their other accomplishments have been overshadowed by the joint legacy of a planet seen with their inner vision long before Neptune's disk was identified in the starry reaches of the night sky.

A Short Glossary of Astronomical Terms Used in the Text

Angular diameter. The apparent diameter of an object, measured by the angle it subtends, and expressed in degrees, minutes, and seconds of arc; typical notation: $10° 31' 59''$.

Aphelion. That point of a planet's or comet's orbit most distant from the Sun—opposed to *perihelion.*

Apsides. Those points in an orbit at which a body is at either the greatest or the least distance from its center of attraction.

Astronomical unit. A unit of length equal to the mean radius of the Earth's orbit, or about 92.9 million miles.

Celestial equator. The great circle in which the plane of the Earth's equator intersects the celestial sphere.

Conjunction. The meeting or passing of two or more celestial bodies in the same degree of the zodiac. Celestial bodies are said to be in conjunction when they are seen in the same part of the heavens, or have the same longitude or right ascension (*q.v.*).

Declination. Angular distance from the celestial equator, measured either northward or southward on an hour or degree circle.

Direct motion. Motion of a heavenly body from west to east, or in the direction of the general planetary motion.

Eccentricity. The ratio of the distance separating the foci of an ellipse to the length of its major axis. In general, the higher the eccentricity, the more compressed the ellipse.

Ecliptic. That great circle of the celestial sphere which is the apparent path of the Sun among the stars, or that of the Earth as seen from the Sun; the intersection of the plane of the Earth's orbit with the celestial sphere.

Elements of a planet's or comet's orbit. The mathematical

quantities that define the form and position of the orbit in space, and express all the circumstances of revolution round the Sun.

Ephemeris. A publication giving the computed places of the celestial bodies for each day of the year, or for other regular intervals. An astronomical almanac.

Epoch. An instant of time or a date selected as a point of reference for which are given values of the data under consideration, which vary with the time.

Equinox. Either of the two points where the celestial equator intersects the ecliptic. The *vernal equinox* is the first point of Aries; the *autumnal equinox,* the first point of Libra.

Geocentric. Pertaining to, or measured from, the Earth's center; having, or relating to, the Earth as a center.

Heliocentric. Pertaining to, or measured from, the Sun's center, or appearing to be seen from it; having, or relating to, the Sun as a center.

Inclination. The angle by which the plane of a planet's or comet's orbit deviates from the plane of the ecliptic.

Latitude. The angular distance of a celestial body from the ecliptic. Celestial latitude may be *geocentric* or *heliocentric;* the latter is sometimes called *true* latitude.

Libration. A real or apparent oscillatory motion; of the Moon, a slow apparent swinging of the visible half of the Moon's surface, causing parts near the limb or edge to be alternately visible and invisible.

Longitude. Angular distance measured along a great circle of reference, from the intersection of the reference circle with an adopted zero meridian to the similar intersection with the meridian passing through the object whose position is being defined. *Celestial longitude* is measured eastward along the ecliptic from the vernal equinox (see *Equinox*) to the great circle passing through the pole of the ecliptic and the object. It may be *geocentric* or *heliocentric.*

Magnitude. The apparent brightness of a celestial body, especially of a star, expressed on a scale of numbers. The brightest stars are represented by the lowest numbers, including zero and negative values. Thus, the magnitude of Sirius, the brightest star, is −1.6.

Mean distance of a planet from the Sun. The mean of the perihelion and aphelion distances (*q.v.*), which is equivalent to half the major axis of the orbit.

Node. A point at which the path of a planet or comet intersects the ecliptic. A node passed as the body goes north is called an *ascending node,* that passed in going south, the *descending node.*

Nutation. A periodic libratory motion of the Earth's axis, like the nodding of a top, by which its inclination to the plane of the ecliptic varies with a range of only a few seconds of arc, so that the celestial poles describe wavy tracks, not circular, around the poles of the ecliptic.

Obliquity of the ecliptic. The slowly varying angle between the planes of the Earth's equator and orbit.

Opposition. The situation of a celestial body with respect to another when their difference in longitude is 180°; especially, such position of a planet or satellite with respect to the Sun. A planet seen in opposition is at its nearest approach to the Earth for that particular cycle.

Parallax. The difference in direction of a heavenly body as seen from two different points. In the measurement of *stellar parallax* the two points are at the opposite ends of a diameter of the Earth's orbit, and the parallax is defined as the angle subtended by half this diameter. In measuring *solar parallax,* from which we can determine the distance of the Earth from the Sun by triangulation, the two points are on opposite sides of the Earth, and the parallax is the angle subtended by the semidiameter of the Earth.

Perihelion. That point of the orbit of a planet or comet which is nearest the Sun—opposed to *aphelion.*

Period. The time in which a planet or satellite completes a single revolution about its primary. The *sidereal period* is measured by using a star seen from the primary or a line joining the primary and a star as a starting and finishing point.

Perturbation. A disturbance of the regular elliptic or other motion of a celestial body, produced by some force additional to that which causes its regular motion; for example, the perturbations of planets are due to their attraction on each other.

Precession of the equinoxes (*q.v.*). A slow westward motion of the equinoctial points along the ecliptic, caused primarily by the action of the Sun and Moon upon the protuberant matter about the Earth's equator in connection with its diurnal rotation.

Radius vector. A straight line joining the center of an attracting body with that of a body describing an orbit around it; for example, a line joining the Sun and a planet.

Retrograde motion. Motion of a heavenly body from east to west, or contrary to the direction of the general planetary motion.

Right ascension. The degree of the equator which, in the right sphere, rises at the same moment with any celestial body; now usually the angle measured eastward or counterclockwise along the celestial equator from the vernal equinox to the hour circle passing through any celestial body. It is expressed in hours, minutes, and seconds; for example, $12^h\ 41^m\ 13^s$.

Secular. Persisting over a long period of time.

Sidereal drive. A mechanism that moves a telescope in synchronism with the apparent motion of the celestial sphere.

Tropic. Either of two small circles of the celestial sphere, on each side of and parallel to the equator, at a distance of 23° 27′, which the Sun reaches at its greatest declination north or south. The northern circle is called the *Tropic of Cancer,* and the southern the *Tropic of Capricorn,* from the two signs at which the tropics touch the ecliptic.

NOTES

Introduction

1. Galileo Galilei, *Siderius nuncius,* p. 21.

Chapter 1. URANUS, 1781–1809

1. Lubbock, *Herschel chronicle,* p. 59.
2. For an explanation of this method, see Otto Struve, "The first stellar parallax determination," pp. 186–190.
3. William Herschel, *Collected scientific papers,* p. xxix.
4. Lubbock, *Herschel chronicle,* p. 80.
5. *Philosophical Transactions of the Royal Society of London 71,* 492–501 (1781).
6. Lubbock, *Herschel chronicle,* p. 86.
7. Fixlmillner, *Berliner Astronomisches Jahrbuch* (1787), p. 249.
8. Rudaux and Vaucouleurs, *Larousse encyclopedia of astronomy,* p. 63.
9. Bode, *Anleitung.*
10. Bode, *Von dem neuen . . . Hauptplaneten,* tr. Shapley and Howarth, p. 181.
11. *Ibid.*
12. *Ibid.,* p. 182.
13. Zach, *Monatliche Correspondenz 3,* 592 (1801).
14. Piazzi, *Della scoperta del nuovo pianeta Cerere Ferdinandea.*
15. *Philosophical Transactions 92,* 227 (1802).
16. Gauss, *Theoria motus,* tr. Davis, p. xiv.

Chapter 2. URANUS, 1809–1840

1. Dunnington, *Gauss,* p. 73.
2. Bouvard, *Tables astronomiques,* p. xiv.
3. Delambre, *Histoire de l'astronomie,* p. 594.
4. Bessel, *Populäre Vorlesungen,* p. 448.
5. Leverrier, abstract of "Recherches sur les mouvements d'Uranus," *Comptes rendus 22,* 907–908 (1846)).
6. Smart, "John Couch Adams and the discovery of Neptune," p. 39.
7. Clairaut, *Journal des sçavans* (Paris, 1759), p. 86.
8. See Hansen's letter reproduced in Gould, *Report on Neptune,* p. 11.
9. Airy, "Account," p. 123.
10. Valz, letter on Halley's comet, *Comptes rendus 1,* 130 (1835).
11. *Astronomische Nachrichten 13,* 94 (Altona, 1836).
12. *Comptes rendus 2,* 154 (1836).
13. *Ibid.,* p. 311.

14. *Ibid., 3,* 141 (1836).
15. Somerville, *Mechanism of the heavens,* p. 399.
16. Somerville, *Connexion of the physical sciences,* ed. 3, p. 74.
17. Airy, "Account," p. 125.
18. Gould, *Report on Neptune,* p. 13.
19. Mädler, *Populäre Astronomie,* p. 345.
20. Bessel, *Populäre Vorlesungen,* p. 449.

Chapter 3. LEVERRIER

1. Bertrand, "Éloge historique," p. 5.
2. U. J. J. Leverrier, *Annales de Chimie et de Physique* (Paris) *60,* 174 (1835); *65,* 257 (1837).
3. Bertrand, "Éloge historique," p. 6.
4. *Ibid.*
5. Villarçeau, "Discours au nom des astronomes de l'Observatoire," *Comptes rendus 85,* 584 (1877).
6. Laplace, *Mécanique céleste,* book 9, chap. 2.

Chapter 4. ADAMS

1. The biographical material in this chapter was drawn from three sources: (1) J. W. L. Glaisher's "Biographical notice," in *The scientific papers of John Couch Adams* (Cambridge, England, 1896), vol. 1, pp. xv–xlviii; (2) W. M. Smart's "John Couch Adams and the discovery of Neptune," *Occasional Notes of the Royal Astronomical Society,* no. 11, vol. 2 (London, 1947); (3) John Couch Adams's notebooks in the library of St. John's College, Cambridge University.
2. British Association, *Report of the first and second meetings,* p. 154. See page 46 above.

Chapter 5. THE HYPOTHESIS

1. Adams, "Elements of the comet of Faye."
2. Airy, "Account," p. 128.
3. *Ibid.* Airy's letter was dated February 15, 1844.
4. Adams, "An explanation of the . . . motion of Uranus," p. 4.
5. Smart, "John Couch Adams," p. 50.
6. Airy, "Account," p. 129.
7. *Ibid.*
8. *Ibid.,* pp. 129–130.
9. Airy, ed., *Autobiography,* p. 3.
10. *Ibid.,* pp. 2–3.
11. Maunder, *Royal Observatory,* p. 118.
12. *Ibid.,* p. 117.
13. Airy, "Account," p. 124.
14. *Ibid.,* p. 126. See page 54 above.
15. *Ibid.,* p. 131.
16. *Ibid.,* p. 128.
17. Smart, "John Couch Adams," p. 56.
18. London *Athenaeum,* December 19, 1846, p. 1300.

19. Jones, *John Couch Adams,* p. 19.
20. Littlewood, *Mathematician's Miscellany,* p. 131.
21. Jones, *John Couch Adams,* p. 22.
22. *Comptes rendus 21,* 1050 (1845).
23. Airy, "Account," p. 131.

CHAPTER 6. THE DISCOVERY

1. *Comptes rendus* 22, 907 (1846).
2. Leverrier, "Recherches," p. 185.
3. Airy, "Account," p. 132.
4. *Ibid.*
5. *Ibid.*, p. 133.
6. *Ibid.*, p. 134.
7. *Ibid.*, p. 135.
8. *Ibid.*, p. 136.
9. Challis, "Special report," p. lii.
10. Airy, "Account," p. 136.
11. *Ibid.*, p. 137.
12. *Ibid.*
13. *Ibid.*, p. 140.
14. Herschel, "Le Verrier's Planet."
15. Galle, *Olai Roemeri,* p. 33.
16. Turner, obituary notice of Galle, p. 278.
17. Joseph von Fraunhofer (1787–1826) was an outstanding optician and experimenter who made some of the finest astronomical instruments in Europe. The Berlin refractor was a near twin to Fraunhofer's great Dorpat telescope, and in 1846 it was surpassed in size by only four other refractors.
18. This map was constructed by Carl Bremiker (1804–1877), an inspector in the Prussian Ministry of Commerce, who also worked on the *Berliner Astronomisches Jahrbuch.* He completed six charts of the Berlin Academy's *Star Atlas.*

CHAPTER 7. NEPTUNE

1. Manuscript letter in the library of the Paris Observatory.
2. Manuscript letter in the library of the Paris Observatory.
3. Manuscript letter in the library of the Paris Observatory.
4. Challis, "Special report," p. liii.
5. Airy, "Account," p. 147.
6. Smart, "John Couch Adams," p. 63.
7. Airy, "Account," p. 147.
8. Challis, "Special report," p. liii.
9. Airy, "Account," p. 143.
10. Turner, obituary notice of Galle, p. 280.
11. Manuscript letter in the library of the Paris Observatory.
12. Arago, "Examen," p. 662.
13. *Ibid.*
14. *Astronomische Nachrichten 25,* 81, 82 (Altona, 1846).

15. Herschel, "Le Verrier's Planet."
16. Challis, Letter to Arago, *Comptes rendus 23,* 715 (1846).
17. Smart, "John Couch Adams," p. 65.
18. *Ibid.*
19. *La Semaine,* Paris, October 25, 1846, p. 794. François Guizot was premier of France under Louis Philippe from 1840 to 1848.
20. Arago, "Examen," p. 749.
21. *Ibid.,* p. 751.
22. *Ibid.,* p. 754.
23. Manuscript letter in the library of the Paris Observatory.
24. Joseph Bertrand, "Sur les erreurs graves."
25. Airy, "Account."
26. Challis, "Account."
27. Adams, *Monthly Notices* 7, 149 (1847).
28. R. A. Sampson, "The decade 1840–1850," in Dreyer and Turner, eds., *History of the Royal Astronomical Society,* p. 96.
29. Smart, "John Couch Adams," p. 70.
30. Maunder, *Royal Observatory Greenwich,* p. 116.
31. Hall, *England and the Orleans monarchy,* p. 393.
32. *Monthly Notices* 7, 216 (1847).
33. Everett, "On the new planet Neptune," *Astronomische Nachrichten 25,* 381, 382 (Altona, 1847).
34. *Ibid.,* pp. 291, 303–306.
35. Peirce, "Investigation," p. 65.
36. Smart, "John Couch Adams," p. 83.
37. Biot, "Sur la planète, nouvellement découverte."

BIBLIOGRAPHY

Primary Sources

Manuscripts

John Couch Adams. Manuscripts on the perturbations of Uranus, 1841–1846. Library of St. John's College, Cambridge University, Cambridge, England.

Urbain Jean Joseph Leverrier. Manuscript of the memoir: "Recherches sur le mouvement de la planète Herschel (dite Uranus)." Library of the Paris Observatory, Paris, France.

Félix Victor Mauvais. Manuscript note on the identification of the planet Neptune with the stars observed by Lalande in May of 1795. Ms C6 (12), library of the Paris Observatory.

Manuscript letters in the library of the Paris Observatory: Airy to Leverrier, October 19, 21, 26, 1846; February 28, 1847. Encke to Leverrier, September 28, 1846. Galle to Leverrier, September 25, 1846. Leverrier to Galle, September 18, 1846; to Encke, October 6, 1846; to Schumacher, October 15, November 25, 1846; to Airy, October 23, 1846. Schumacher to Leverrier, September 28, September 29, October 8, 1846, January 19, 1847; to Arago, November 6, 1846.

Printed Matter

Adams, John Couch, "An explanation of the observed irregularities in the motion of Uranus, on the hypothesis of disturbances caused by a more distant planet; with a determination of the mass, orbit, and position of the disturbing body," *Appendix to the Nautical Almanac for the Year 1851* (London, 1846). *Monthly Notices of the Royal Astronomical Society* 7, 149–152 (London, 1847).

——— "Elements of the comet of Faye, computed by J. C. Adams, Esq. of St. John's College, Cambridge," *Monthly*

Notices of the Royal Astronomical Society 6, 20–21 (London, 1844).

——— *Lectures on the lunar theory* (Cambridge, England, 1900).

——— Letter on de Vico's comet, *Times* (London), October 15, 1844, p. 8.

Adams, William Grylls, ed., *The scientific papers of John Couch Adams* (2 vols.; Cambridge, England, 1896).

Airy, George Biddell, "Account of some circumstances historically connected with the discovery of the planet exterior to *Uranus*," *Monthly Notices of the Royal Astronomical Society* 7 (No. 9), 121–144 (London, 1846).

——— *Gravitation: An elementary explanation of the principal perturbations in the solar system* (London, 1834).

——— *Mathematical tracts on the lunar and planetary theories, the figure of the Earth, precession and nutation, the calculus of variations and the undulatory theory of optics* (Cambridge, England, 1842).

——— "Mr. Adams and the new planet" (letter), *Athenaeum* (London), February 27, 1847, p. 229.

——— "Name of the new planet" (letter), *Athenaeum* (London), February 20, 1847, pp. 199–200.

——— *Reduction of the observations of planets made at the Royal Observatory, Greenwich, from 1750 to 1830* (London, 1845).

——— *Six lectures on astronomy* (London, 1848).

——— and Urbain Jean Joseph Leverrier, "Nouvelle détermination de la différence de longitude entre les observatoires de Paris et de Greenwich," *Comptes rendus 39,* 553–566 (Paris, 1854).

Airy, Wilfrid, ed., *Autobiography of Sir George Biddell Airy* (Cambridge, England, 1896).

Arago, François, "Examen des remarques critiques et des questions de priorité que la découverte de M. Le Verrier a soulevées," *Comptes rendus 23,* 751–754 (Paris, 1846).

Bertrand, Joseph, "Sur les erreurs graves commises par un géomètre étranger," *Comptes rendus 23,* 827–832 (Paris, 1846).

Bessel, Friedrich Wilhelm, *Fundamenta astronomiae pro A. 1755 deducta ex observationibus viri incomparibilis James*

Bradley in specula astronomica Grenovicensi per A. 1750–62 institutis (Regiomonti, 1818).

——— *Populäre Vorlesungen* (Hamburg, 1848).

Biot, Jean Baptiste, "Sur la planète, nouvellement découverte par M. Le Verrier, comme conséquence de la théorie de l'attraction," *Journal des Savants 12,* series 3, 64–85 (Paris, 1847).

Bode, Johann Elert, *Anleitung zur Kenntniss des gestirnen Himmels* (ed. 2, Hamburg, 1772).

——— *Von dem neu entdecken Planeten* (Berlin, 1784).

——— *Von dem neuen, zwischen Mars und Jupiter entdeckten achten Hauptplaneten des Sonnensystems* (Berlin, 1802); tr. H. Shapley and H. E. Howarth, *A source book in astronomy* (New York, 1929), pp. 180–182.

Bonnet, Charles, *Contemplation de la nature* (ed. 2, tr. Johann Titius, Wittenberg, 1772).

Bouvard, Alexis, *Tables astronomiques publiées par le Bureau des Longitudes de France contenant les Tables de Jupiter, de Saturne et d'Uranus, construites d'après la théorie de la Mécanique céleste* (Paris, 1821).

Brewster, David, *A treatise on optics* (London, 1831).

Challis, James, "Account of observations at the Cambridge Observatory for detecting the planet exterior to Uranus," *Monthly Notices of the Royal Astronomical Society 7,* 145–149 (London, 1846).

——— letter to François Arago, *Comptes rendus 23,* 715–716 (Paris, 1846).

——— letter to the editor of the *Athenaeum, Athenaeum,* December 19, 1846, 1300 (London).

——— "Special report of proceedings in the Observatory relative to the new planet" (December 12, 1846), in W. G. Adams, ed., *The scientific papers of John Couch Adams* (Cambridge, England, 1896), vol. 1, pp. xlix–liv.

Clairaut, Alexis Claude, "Mémoire sur la comète de 1682 adressée à MM. les auteurs de Journal des sçavans par M. Clairaut," *Journal des sçavans,* 1759, 80–96 (Paris).

Delambre, Jean Baptiste Joseph, *Histoire de l'astronomie au dix-huitième siècle* (Paris, 1827).

——— *Tableaux des éléments des comètes, astronomie théorique et pratique* (Paris, 1814), vol. 3.

Galilei, Galileo, *Siderius nuncius,* in Stillman Drake, ed., *Discoveries and opinions of Galileo* (Garden City, New York: Doubleday, 1957).

Galle, Johann Gottfried, *Olai Roemeri triduum observationum astronomicarum a. 1706 institutarum reductum et cum tabulis comparatum* (Berlin, 1845).

——— "Ueber die erste Auffindung des Planeten Neptun," *Copernicus 2,* 96–97 (Dublin, 1882).

Gauss, Karl Friedrich, *Theoria motus corporum coelestium in sectionibus conicis solem ambientum* (Hamburg, 1809), tr. C. H. Davis, *Theory of the motion of the heavenly bodies moving about the sun in conic sections* (Boston, 1857).

Herschel, John Frederick William, letter on "Le Verrier's planet," *Athenaeum* (London), October 3, 1846, p. 1019.

——— *Outlines of astronomy* (Philadelphia, 1849).

Herschel, Mrs. John, *Memoir and correspondence of Caroline Herschel* (New York, 1876).

Herschel, William, "Account of a comet," *Philosophical Transactions of the Royal Society* (London) *71,* 492–501 (1781).

——— *Collected scientific papers* (2 vols.; London, 1912).

Hind, John Russell, "Discovery of Le Verrier's planet" (letter), *Times* (London), October 1, 1846, p. 8.

Kant, Immanuel, *Allgemeine Naturgeschichte* (Frankfort, 1797).

Kepler, Johann, *Harmonices mundi libri V* (Linz, 1619), vol. 3, op. 3.

——— *Opera omnia* (Frankfurt, 1859), vol. 1.

Lalande, Joseph Jérôme Le Français de, *Astronomie* (ed. 3, Paris, 1792), vol. 1.

——— *Histoire céleste française* (Paris, 1801).

Laplace, Pierre Simon de, *Traité de mécanique céleste* (5 vols.; Paris, 1799–1825).

Lassell, William, letter on a satellite of Neptune, *Times* (London), October 14, 1846, p. 7.

Laugier, Paul Auguste Ernest, "Mémoire sur quelques comètes anciennes," *Comptes rendus* 22, 148–156 (Paris, 1846).

Leverrier, Urbain Jean Joseph, "Calcul de la valeur des perturbations que la comète découverte par De Vico, 1844, Aug. 22, peut éprouver par l'action de la Terre," *Comptes rendus 19,* 666–670 (Paris, 1845).

——— "Comparaison des observations de la nouvelle planète avec la théorie déduite des perturbations d'Uranus," *Comptes rendus 23,* 771 (Paris, 1846).

——— *Dévelopements sur plusiers points de la théorie du perturbations des planètes* (3 parts; Paris, 1841, 1842).

——— *Examen de la discussion soulevée au sein de l'Académie des Sciences au sujet de la découverte de l'attraction universelle* (Paris, 1869).

——— "Premier mémoire sur la théorie d'Uranus," *Comptes rendus 21,* 1050–1055 (Paris, 1845).

——— "Recherches sur le mouvement de la planète Herschel (dite Uranus)," *Connaissance des Temps pour 1849, Additions* (Paris, 1846), pp. 3–254.

——— "Recherches sur les mouvements d'Uranus," *Comptes rendus 22,* 907–918 (Paris, 1846).

——— "Seconde note sur les perturbations de la planète Uranus," *Comptes rendus 14,* 660–663 (Paris, 1842).

——— "Sur la comète observée de M. Faye, 1843, Nov. 22 et sur son identité avec la comète de Lexell," *Comptes rendus 18,* 826–827 (Paris, 1844).

——— "Sur la détermination des inégalités séculaires des planètes," *Connaissance des temps for 1844,* 28–110 (Paris, 1841).

——— *Sur la planète Neptune* (Paris, 1848).

——— "Sur la planète qui produit les anomalies observées dans le mouvement d'Uranus—Détermination de sa masse, de son orbite et de sa position actuelle," *Comptes rendus 23,* 428–438, 657–662 (Paris, 1846).

——— "Sur les variations séculaires des orbites des planètes," *Comptes rendus 9,* 370–374 (Paris, 1839).

——— "Sur les variations séculaires des orbites planetaires," *Connaissance des temps pour 1843, Additions* (Paris, 1840), p. 24.

——— *Théorie nouvelle du mouvement de la planète Neptune* (Paris, 1874).

Lexell, Anders Johann, circular elements of Uranus, *Acta Academiae Scientarium Imperialis Petropolitanae,* Mémoires, v. 4, p. 307 (St. Petersburg, 1780).

Mädler, Johann Heinrich von, *Populäre Astronomie* (Berlin, 1841).

Mayer, Johann Tobias, *Opera inedita,* vol. 1, ed. G. Ch. Lichtenberg (Göttingen, 1775).

Peirce, Benjamin, "Formulae in the theory of Neptune" (read December 7, 1847), *Proceedings of the American Academy of Arts and Sciences 1* (1846–1848), 287–295 (Boston, 1848).

——— "Investigation into the action of Neptune to Uranus," *ibid.,* pp. 65–95.

——— "Perturbations of Uranus by Neptune," *Proceedings of the American Philosophical Society 5,* 15–16 (Philadelphia, 1848).

Petersen, A. C., "Nachsuchung früherer Beobachtungen des *Le Verrier*'schen Planeten," *Astronomische Nachrichten 25,* 291, 303–306 (Altona, 1847).

Piazzi, Giuseppe, *Della scoperta del nuova pianeta Cerere Ferdinandea* (Palermo, 1802).

Schumacher, Heinrich Christian, circular announcing the discovery of the planet Neptune; letter from J. F. Encke (Altona, September 29, 1846).

Sheepshanks, (Rev.) Richard, *A reply to Mr. Babbage's letter to "The Times," "On the planet Neptune and the Royal Astronomical Society's medal"* (London, 1847).

Somerville, Mrs. Mary Fairfax, *The connexion of the physical sciences* (ed. 1, London, 1834; ed. 3, rev., London, 1836).

——— *The mechanism of the heavens* (London, 1831).

Struve, Friedrich Georg Wilhelm, "On the denomination of the planet newly discovered beyond the orbit of Uranus" (Polkowa, 17 (29) December 1846), tr. in *Athenaeum* (London), February 20, 1847, pp. 199–200.

——— *Ueber den neuen Hauptplaneten Neptun* (Saint Petersburg, 1846).

Struve, Otto Wilhelm, "Bemerkungen über die gegen Herrn Le Verrier erhobenen Angriffe in Betreff der Identität des Neptun mit dem Planeten, dessen Ort er aus den Uranusstörungen theoretisch abgeleitet hatte" (read October 20, 1848), *Bulletin de la classe physico-mathématique de l'Académie des Sciences de Saint-Pétersbourg 7,* col. 321–336 (Saint-Petersburg, 1849).

Valz, Jean Élix Benjamin, Éléments provisoires de la planète de M. Le Verrier pour l'époque du 7 décembre 1846" *Comptes rendus 24,* 33–35 (Paris, 1847).

——— letter on Halley's comet, *Comptes rendus 1,* 130–131 (Paris, 1835).
——— "Sur la comète découverte par M. Faye le 22 novembre 1843," *Comptes rendus 18,* 764–767 (Paris, 1844).
Walker, Sears Cook, "Elements of the planet Neptune" (letter), *Proceedings of the American Philosophical Society 4,* 332–335 (Philadelphia, 1847).
——— "Elliptic elements of the planet Neptune," *Proceedings of the American Philosophical Society 4,* 378 (Philadelphia, 1847).
——— "Ephemeris of the planet Neptune," *Proceedings of the American Philosophical Society 5,* 20–21 (Philadelphia, 1848).
——— "Investigations which led to the detection of the coincidence between the computed place of the planet Leverrier and the observed place of a star recorded by Lalande in May, 1795" (read February 13th, 1847), *Transactions of the American Philosophical Society* (n.s.) *10,* 141–153 (Philadelphia, 1853).
——— *Researches relative to the planet Neptune* (Smithsonian Institution Publication 3; Washington, D. C., 1849).
Zach, Franz Xaver von, "Über einen zwischen Mars und Jupiter längst vermutheten, nun wahrscheinlich entdeckten neuen Hauptplaneten unseres Sonnen-Systems," *Monatliche Correspondenz zur Beförderung der Erd und Himmelskunde 3,* 592–623 (Gotha, 1801).

Secondary Sources

Books

Abetti, Giorgio, *The history of astronomy* (New York: H. Schuman, 1952); tr. by B. B. Abetti, from *Storia dell'astronomia.*
Andrade, E. N. da C., *Sir Isaac Newton* (Garden City, New York: Doubleday, 1958).
Arago, François, *Astronomie populaire* (Paris, 1857).
Armitage, Angus, *A century of astronomy* (London: Low, 1950).
Astronomischer Jahresbericht, vols. 1–58 (Berlin, 1899–1958).

Berry, Arthur, *A short history of astronomy* (New York, 1898).
Brisse, M. R., ed., Catalogue of the exposition: *Leverrier et son temps* (Paris, 1946).
British Association for the Advancement of Science, *Report of the first and second meetings* (London, 1833).
Busco, Pierre, *L'Évolution de l'astronomie au XIXe siècle* (Paris, 1912).
Cecil, Algernon, *British foreign secretaries 1807–1916* (London, 1927).
Clerke, Agnes M., *A popular history of astronomy during the nineteenth century* (ed. 4, London, 1902).
Cohen, I. Bernard, *The birth of a new physics* (Garden City, New York: Doubleday, 1960).
Doig, Peter, *A concise history of astronomy* (London: Chapman and Hall, 1950).
Dreyer, J. L. E., *A history of astronomy from Thales to Kepler* (New York: Dover, 1953).
——— and H. H. Turner, eds., *History of the Royal Astronomical Society 1820–1920* (London, 1923).
Dunnington, G. Waldo, *Carl Friedrich Gauss: Titan of science* (New York: Exposition Press, 1955).
Flammarion, Camille, *Astronomie populaire* (Paris, 1880).
Gooch, G. P., and A. W. Ward, eds., *The Cambridge history of British foreign policy, 1783–1919* (3 vols.; Cambridge, England, 1922, 1923).
Gould, Benjamin Apthorp, Jr., *Report on the history of the discovery of Neptune* (Washington, D. C., 1850).
Grant, Robert, *History of physical astronomy* (London, 1852).
Hall, John, *England and the Orleans monarchy* (London, 1912).
Hoefer, Ferdinand, *Histoire de l'astronomie* (Paris, 1873).
Houghton, Walter E., *The Victorian frame of mind, 1830–1870* (New Haven: Yale University Press, 1957).
Howarth, J. R., *The British Association for the Advancement of Science: A retrospect 1831–1921* (London, 1922).
Jones, Harold Spencer, *John Couch Adams and the discovery of Neptune* (Cambridge, England: Cambridge University Press, 1947).
King, Henry D., *The history of the telescope* (Cambridge, Massachusetts: Sky Publishing Corporation, 1955).

BIBLIOGRAPHY

Kuhn, Thomas S., *The Copernican revolution* (Cambridge, Massachusetts: Harvard University Press, 1957).

Liais, E., *L'Histoire de la découverte de la planète Neptune* (Leipzig, 1892).

Littlewood, J. E., *A mathematician's miscellany* (London: Methuen, 1957).

Lubbock, Constance A., ed., *The Herschel chronicle* (Cambridge, England, 1933).

Lubbock, John, *Fifty years of science (1831–1881)* (London, 1890).

Mädler, Johann Heinrich von, *Reden und Abhandlungen über gegenstände der Himmelskunde (Die Entdeckung des Neptuns)* (Berlin, 1870).

Maunder, E. Walter, *The Royal Observatory Greenwich* (London, 1900).

McKie, Douglas, *Antoine Lavoisier* (New York: H. Schuman, 1952).

Mitchel, O. M., *The planetary and stellar worlds* (New York, 1848).

Murray, Robert H., *Science and scientists in the nineteenth century* (London, 1925).

Newcomb, Simon, *An investigation of the orbit of Neptune with general tables of its motion* (Smithsonian Contributions to Knowledge, No. 199; Washington, D. C., 1865).

——— *Popular astronomy* (New York, 1878).

——— *The reminiscences of an astronomer* (Boston, 1903).

Newman, James R., ed., *The world of mathematics* (4 vols.; New York: Simon and Schuster, 1956).

Nichol, J. P., *The planet Neptune: An exposition and history* (Edinburgh, 1848).

Pledge, H. T., *Science since 1500* (New York, Harper & Brothers, 1959).

Royal Society of London, *The record of the Royal Society* (London, 1912).

Rudaux, Lucien, and G. de Vaucouleurs, *Larousse encyclopedia of astronomy* (New York: Prometheus Press, 1959).

Sampson, R. A., *A description of Adams's manuscripts on the perturbations of Uranus* (London, 1901).

Shapley, Harlow, and Helen E. Howarth, *A source book in astronomy* (New York, 1929).

Smart, W. M., *Celestial mechanics* (London: Longmans, Green, 1953).

Smyth, W. H., *The cycle of celestial objects continued at the Hartwell Observatory to 1859* (London, 1860).

Somervell, D. C., *English thought in the nineteenth century* (London, 1936).

Teichman, Oskar, *The Cambridge undergraduate 100 years ago* (Cambridge, England, 1926).

Thomson, David, *England in the nineteenth century (1815–1914)* (Harmondsworth: Penguin Books, 1950).

Tisserand, Félix, *Traité de mécanique céleste* (Paris, 1889).

Turner, Herbert Hall, *Astronomical discovery* (London, 1904).

Vaucouleurs, Gerard de, *Discovery of the universe,* tr. B. Pagel (London: Faber and Faber, 1957).

Venn, J. A., ed., *Alumni Cantabrigiensis,* part 2, vol. 4 (Cambridge, England, 1951).

Vincent, M., *Le Mystère de la dècouverte de Neptune par Leverrier* (Paris: Fischbacher, 1958) .

Wackerbarth, A.–T.–D., *Om planeten Neptunus* (Uppsala, 1865).

Williams, Henry Smith, *The great astronomers* (New York, 1930).

Wolf, A., *A history of science, technology, and philosophy in the 16th and 17th centuries* (ed. 2, London: Allen & Unwin, 1950).

—— *A history of science, technology, and philosophy in the eighteenth century* (ed. 2, London: Allen & Unwin, 1952).

Wolf, John B., *France: 1815 to the present* (New York: Prentice-Hall, 1940).

Wolf, Rudolf, *Geschichte der Astronomie* (Munich, 1877).

Articles

Allen, Edward, letter on "The discovery of Neptune," *The Observatory 15,* 260–261 (London, 1892).

"An astronomical might-have-been," *Journal of the British Astronomical Association 61,* 81–82 (Oxford, 1951).

Aoust, Louis (l'abbé), "Le Verrier, sa vie, ses travaux," *Mémoires de l'Académie des Sciences, Belles-Lettres et Arts de Marseille,* 1877–1878, 164–209 (Marseilles, 1878).

BIBLIOGRAPHY

Bertrand, M. J., "Éloge historique de Urbain-Jean-Joseph Leverrier," *Annales de l'Observatoire de Paris, Mémoires 15*, 3–22 (Paris, 1880).

Brasch, F. E., "Sir William Herschel, 1738–1822," *Popular Astronomy 47* (No. 2, February), 1–7 (Northfield, Minnesota, 1939).

Brewster, David, "Researches respecting the new planet Neptune," *North British Review 7* (May), 110–132 (Edinburgh, 1847).

Büdeler, W., "Die Entdeckung des Planeten Neptun," *Neue Physikalische Blätter 2*, 108–109 (Wurtemberg/Baden, 1946).

Danjon, André, "La découverte de Neptune," *Bulletin de la Société Astronomique de France 60*, 255–278 (Paris, 1946).

——— "Le centenaire de la découverte de Neptune," *Ciel et Terre 62*, 369–383 (Bruxelles, 1946).

Dawson, B. H., "Centenario del descubrimiento de Neptuno," *Revista Astronomica 18*, 230–232 (Buenos Aires, 1946).

de Broglie, Louis, "La découverte de Neptune et la science moderne," *Bulletin de la Société Astronomique de France 60*, 246–254 (Paris, 1946).

"De honderdste Varjaring van de Ontdekking van Neptunus," *Heelal 2*, 71–74 (Antwerp, 1946).

Dreyer, J. L. E., "Historical note concerning the discovery of Neptune," *Copernicus 2*, 63–64 (Dublin, 1882).

Eklöf, O., "Neptunus upptäckt—ett hundraårsminne," *Populär Astronomisk Tidsskrift 27*, 383–388 (Stockholm, 1946).

Everett, Edward, "On the new planet Neptune," *Astronomische Nachrichten 25*, 375–388 (Altona, 1847).

Febrer, J., "Sintesis del problema resuelto por Le Verrier," *Urania 31*, 1–7 (Barcelona, 1946).

Fleckenstein, J. O., "Zum 100 Jahrestag der Entdeckung des Planeten Neptun am 23 September, 1846," *Experientia 2*, 1–7 (Basel, 1946).

Gaillot, Jean Baptiste Aimable, "Le Verrier et son oeuvre," *Bulletin des Sciences mathématiques et astronomiques* (2nd series) *2*, 29–40 (Paris, 1878).

Gingerich, Owen, "The naming of Uranus and Neptune," *Astronomical Society of the Pacific Leaflet 352*, 1–7 (San Francisco, 1958).

Glaisher, J. W. L., "Biographical notice" of John Couch Adams, in W. G. Adams, ed., *The scientific papers of John Couch Adams* (Cambridge, England, 1896), vol. 1, pp. xv–xlviii.

——— obituary notice of John Couch Adams, *Monthly Notices of the Royal Astronomical Society 53* (No. 4), 184–209 (London, 1893).

Gondolatsch, F., "Ein astronomischer Gedenktag: Die Entdeckung des Planeten Neptun vor 100 Jahren am 23 September, 1846," *Astronomischer Kalender für das Jahr 1946,* 60–63 (Heidelberg, 1945).

Gould, Benjamin Apthorp, Jr., editor's note, *Astronomical Journal 1,* 78 (Cambridge, Massachusetts, 1851).

Hind, John Russell, obituary notice of Urbain Jean Joseph Leverrier, *Monthly Notices of the Royal Astronomical Society 38* (No. 4), 155–166 (London, 1878).

——— letter on Lamont's 3rd observation of Neptune, *Monthly Notices of the Royal Astronomical Society 11,* 11 (London, 1851).

——— "Unnoticed observations of Neptune," *Monthly Notices of the Royal Astronomical Society 10,* 42 (London, 1850).

Holden, Edward S., "Historical note relating to the search for the planet Neptune in England in 1845–6," *Astronomy and Astro-Physics 11,* 287–288 (Northfield, Minnesota, 1892).

"100 Jahre seit der Entdeckung Neptuns," *Orion 13,* 261 (Genève, 1946).

Jackson, J., "The discoveries of Neptune and Pluto," *The Observatory 75,* 126–127 (Oxford, 1955).

——— "The discovery of Neptune—a defence of Challis," *Monthly Notes of the Astronomical Society of South Africa 8,* 88–89 (Cape Town, 1949).

Jones, H. S., "G. B. Airy and the discovery of Neptune," *Popular Astronomy 55,* 312–316 (Northfield, Minnesota, 1947).

Kuhn, R., "Die Entdeckung des Uranus," *Sternenwelt 1,* 57–59 (München, 1949).

Larink, J., "Hundert Jahre Neptun," *Die Himmelswelt 55,* 1–5 (Bonn, 1947).

Lynn, W. T., "The jubilee of the discovery of Neptune," *The Observatory 19,* 364–366 (London, 1896).

——— "Lamont's observations of Neptune," *The Observatory 18*, 275–276 (London, 1895).

Lyttleton, R. A., "A short method for the discovery of Neptune," *Monthly Notices of the Royal Astronomical Society 118*, 551–559 (London, 1958).

Michkovitch, V. V., "La part de la mécanique céleste et le rôle du hassard dans les découvertes des deux dernières grosses planètes," *Annuaire de notre ciel pour l'an 1952: Observatoire Astronomique de l'Académie Serbe Scientifique 17*, 208–233 (Belgrade, 1951).

"Name des neuen Planeten," *Astronomische Nachrichten 25*, 81–82 (Altona, 1846).

"Opdagelsen af planeten Neptun," *Naturen 71*, 75–76 (Bergen, 1946).

Petersen, A. C., letter to the editor of the *Astronomical Journal 1*, 47–48 (Cambridge, Massachusetts, 1851).

Rines, David, "The discovery of the planet Neptune," *Popular Astronomy 20*, 482–498 (Northfield, Minnesota, 1912).

Royal Astronomical Society, catalogue and program: "Centenary of the discovery of the planet Neptune 23 September 1846" (London, October 8, 1946).

Royal Astronomical Society, "Report of the council to the twenty-seventh annual general meeting," *Monthly Notices of the Royal Astronomical Society 7*, 193–237 (London, 1847).

Silva, G., "Nel centenario della scoperta di Nettuno," *Osservatorio astronomico di Padova, Pubblicazione No. 90*, 13 (Padua, 1947).

Smart, W. M., "John Couch Adams and the discovery of Neptune," *Occasional Notes* of the Royal Astronomical Society, No. 11, Vol. 2, 33–88 (London, 1947).

Stephan, M., "M. Le Verrier, fondateur du nouvel Observatoire de Marseille," *Mémoires de l'Académie des Sciences, Belles-Lettres et Arts de Marseille*, 1879–1880, 1–28 (Marseilles, 1880).

Struve, Otto, "The first stellar parallax determination," in Herbert M. Evans, ed., *Men and Moments in the History of Science* (Seattle, Washington: University of Washington Press, 1959), pp. 177–206.

"The telescopic discovery of Neptune," *Journal of the British Astronomical Association 61*, 166–167 (Oxford, 1951).

Tisserand, Félix, "Les travaux de M. Le Verrier," *La Revue Scientifique de la France et de l'Étranger,* 2nd series, No. 20, November 17, 1877, 457–465 (Paris, 1877).

Turner, Herbert Hall, obituary notice of Johann Gottfried Galle. *Monthly Notices of the Royal Astronomical Society 71,* 275–281 (London, 1911).

Villarçeau, Yvon, "Discours au nom des Astronomes de l'Observatoire," *Comptes rendus 85,* 584–586 (Paris, 1877).

Wattenberg, D., "Die Entdeckung des Planeten Neptun auf der Berliner Sternwarte vor 100 Jahren," *Die Naturwissenschaften 33,* 129–132 (Berlin, 1946).

——— "Hundert Jahre Planet Neptun," *Zeitschrift für Naturforschung 1,* 540–543 (Wiesbaden, 1946).

Other Periodicals

Athenaeum, October 3, 15, 17; November 21; December 19, 1846. February 20, 1947. London.

Le Charivari, October 10, 12, 17, 24; December 4, 1846. Page 3 of the December issue was devoted to a full-page lithograph by Honoré Daumier, entitled, "Les Bons bourgeois, No. 55; Recherche infructeuse de la planète Leverrier." Paris.

Le Commerce, October 14, 1846. Paris.

Le Constitutionnel, October 9, 1846. Paris.

L'Époque, September 10; October 7, 26; November 7, 1846. Paris.

L'Esprit Public, October 13, 16, 18, 21; November 8, 1846. Paris.

La Gazette de France, October 8, 11, 1846. Paris.

Les Guêpes, January, 1847. Paris.

L'Illustration, October 10, 17; November 7, 28, 1846. Paris.

London *Times,* October 15, 1844; October 1, 14, 1846; March 15, 1847. London.

Le Moniteur Universel, October 10, 18, 19, 1846. Paris.

Le National, September 30; October 6, 7, 8, 14, 21, 22, 23, 28; November 3, 10, 25; December 20, 1846. Paris.

La Patrie, October 27, 1846. Paris.

La Presse, October 8, 1846; September 11, 25, 1848. Paris.

La Réforme, September 13; October 13, 25, 1846. Paris.

La Revue Britannique, October 1846. Paris.

Revue des Deux-Mondes, October 15, 1846; January 15; October 15, 1848. Paris.

BIBLIOGRAPHY

La Semaine, October 4, 11, 18, 25, 1846. Paris.
Le Siècle, October 7, 1846. Paris.
La Silhouette, October 11; November 1, 29, 1846. Paris.
L'Univers, October 4, 11, 21; November 8, 1846. Paris.

INDEX

The more important page references are in italic type

INDEX

INDEX

INDEX